The
Great Ape Project

The
Great Ape Project

Equality beyond Humanity

Edited by Paola Cavalieri
and Peter Singer

ST. MARTIN'S PRESS
NEW YORK

Library of Congress Cataloging-in-Publication Data

The Great ape project : equality beyond humanity / edited by Peter Singer.
p.   cm.
ISBN 0-312-10473-1
1. Animal rights.   2. Animal rights movement.   I. Singer, Peter.
HV4711.G73   1994
179'.3—dc20   93-35892   CIP

First published in Great Britain by Fourth Estate Limited.

First U.S. Edition: January 1994
10 9 8 7 6 5 4 3 2 1

Contents

Preface

We are human, and we are also great apes. Our membership of the human species gives us a precious moral status: inclusion within the sphere of moral equality. Those within this sphere we regard as entitled to special moral protection. There are things that we may not do to them. They have basic rights that are denied to those outside this sphere. This book urges that in drawing the boundary of this sphere of moral equality, we should focus not on the fact that we are human beings, but rather on the fact that we are intelligent beings with a rich and varied social and emotional life. These are qualities that we share not only with our fellow humans, but also with our fellow great apes. Therefore, we should make membership of this larger group sufficient entitlement for inclusion within the sphere of moral equality. We seek an extension of equality that will embrace not only our own species, but also the species that are our closest relatives and that most resemble us in their capacities and their ways of living.

This step is a cautious one: there are relatively few great apes in the world, and to extend equality to them would require a much more modest rearrangement of our lives than, say, the extension of equality to all mammals. Some people, among them some of the contributors to this book, would like to see a much larger extension of the moral community, so that it includes a wider range of nonhuman animals. Our epilogue suggests one way in which this project might have a significance that is wider than its immediate aim. This immediate aim is, though limited, still a step of true historical importance. As the essays in this book demonstrate, we now have sufficient information about the capacities of chimpanzees, gorillas and orang-utans to make it clear that the moral boundary we draw between us and them is indefensible. Hence the time is ripe for extending full moral equality to members of other species, and the case for so doing is overwhelming.

For support in our project we wrote not only to people who know nonhuman apes well, from long periods of observing them or from communicating with them, but also to academics from many different disciplines. The chapters that follow show that we have gathered an enthusiastic group of scholars in both the sciences and the humanities, committed to the single goal of including the nonhuman great apes within the moral community.

The essays are written from a variety of perspectives – including anthropology, psychology, ethology and ethics – but together they form a single, challenging picture. Visions of the apes living free in their native forests contrast with descriptions of the miserable lives that many great apes are now forced to live under human tyranny, whether in zoos, laboratories or other captive conditions. Several essays attempt to understand not only the biological nature of the great apes, but also their own point of view. Others consider how we humans have sought to emphasise the differences between us and them, and to play down the similarities. Ethics appears on the scene, putting the data about apes together with our own rhetoric about the basis of equality between humans. This points towards the goal of this entire enterprise: a reassessment of the moral status of chimpanzees, gorillas and orang-utans and the acceptance of some nonhuman animals as persons. This would be a far-reaching change, as the essays in the final section make clear.

The core of this book is an encounter between ethology and ethics. These two disciplines are linked by the Greek word *ethos*, which means customs or habits. The link between the two fields has almost been obliterated by the fact that ethology is usually understood as the study of animal behaviour, while ethics is commonly limited to human beings. Moreover, in ethology the root word is taken in a descriptive sense, so that the focus is on how animals *do* act, whereas in ethics the emphasis is normative, on how we *ought* to act. In previous interactions between ethology and ethics, it has been the former that has tried to absorb the latter, by producing some kind of scientific, or more specifically evolutionary, basis for ethics. That attempt is often used as a basis for very conservative ethical conclusions. In the present volume, however, the opposite has happened: ethics subsumes ethology and from this emerges a highly innovative social project. At a more general level, we see this project as part of a process in which we restore ethics to its proper place in determining the shape of the society in which we live.

On the pages immediately following this preface we print the *Declaration on Great Apes* that sets out the immediate objective of this book, and to which all contributors subscribe. If you endorse this Declaration, please write to us indicating your support. Thus we shall build an international coalition that will have as its sole objective the inclusion of chimpanzees, gorillas and orang-utans within the community of equals. At a theoretical level this will mean a recognition that the limits of our moral community should not run parallel to the boundary of our species. At a practical level, it will entail freeing all imprisoned chimpanzees, gorillas and orang-utans and returning them to an environment that accords with their physical, mental and social needs. Royalties earned from the book will go towards this objective. In order to maximise the amount available, all the contributors have agreed to forgo payment for their contributions. You may write to: The Great Ape Project, PO Box 1023, Collingwood, Melbourne, Victoria, Australia 3066; or Fax to International + 39 2 481 4784.

Paola Cavalieri
Peter Singer

A Declaration on Great Apes

We demand the extension of the community of equals to include all great apes: human beings, chimpanzees, gorillas and orang-utans.

'The community of equals' is the moral community within which we accept certain basic moral principles or rights as governing our relations with each other and enforceable at law. Among these principles or rights are the following:

1 **The Right to Life**
   The lives of members of the community of equals are to be protected. Members of the community of equals may not be killed except in very strictly defined circumstances, for example, self-defence.

2 **The Protection of Individual Liberty**
   Members of the community of equals are not to be arbitrarily deprived of their liberty; if they should be imprisoned without due legal process, they have the right to immediate release. The detention of those who have not been convicted of any crime, or of those who are not criminally liable, should be allowed only where it can be shown to be for their own good, or necessary to protect the public from a member of the community who would clearly be a danger to others if at liberty. In such cases, members of the community of equals must have the right to appeal, either directly or, if they lack the relevant capacity, through an advocate, to a judicial tribunal.

3 **The Prohibition of Torture**
   The deliberate infliction of severe pain on a member of the community of equals, either wantonly or for an alleged benefit to others, is regarded as torture, and is wrong.

At present, only members of the species *Homo sapiens* are regarded as members of the community of equals. The inclusion, for the first time, of nonhuman animals into this community is an ambitious project. The chimpanzee (including in this term both *Pan troglodytes* and the pygmy chimpanzee, *Pan paniscus*), the gorilla, *Gorilla gorilla*, and the orang-utan, *Pongo pygmaeus*, are the closest relatives of our species. They also have mental capacities and an emotional life sufficient to justify inclusion within the community of equals. To the objection that chimpanzees, gorillas and orang-utans will be unable to defend their own claims within the community, we respond that human guardians should safeguard their interests and rights, in the same ways as the interests of young or intellectually disabled members of our own species are safeguarded.

Our request comes at a special moment in history. Never before has our dominion over other animals been so pervasive and systematic. Yet this is also the moment when, within that very Western civilisation that has so inexorably extended this dominion, a rational ethic has emerged challenging the moral significance of membership of our own species. This challenge seeks equal consideration for the interests of all animals, human and nonhuman. It has given rise to a political movement, still fluid but growing. The slow but steady widening of the scope of the golden rule – 'treat others as you would have them treat you' – has now resumed its course. The notion of 'us' as opposed to 'the other', which, like a more and more abstract silhouette, assumed in the course of centuries the contours of the boundaries of the tribe, of the nation, of the race, of the human species, and which for a time the species barrier had congealed and stiffened, has again become something alive, ready for further change.

The Great Ape Project aims at taking just one step in this process of extending the community of equals. We shall provide ethical argument, based on scientific evidence about the capacities of chimpanzees, gorillas and orang-utans, for taking this step. Whether this step should also be the first of many others is not for The Great Ape Project to say. No doubt some of us, speaking individually, would want to extend the community of equals to many other animals as well; others may consider that extending the community to include all great apes is as far as we should go at present. We leave the consideration of that question for another occasion.

We have not forgotten that we live in a world in which, for at least three-quarters of the human population, the idea of human rights is no more than rhetoric, and not a reality in everyday life. In such a world, the idea of equality for nonhuman animals, even for those disquieting doubles of ours, the other great apes, may not be received with much favour. We recognise, and deplore, the fact that all over the world human beings are living without basic rights or even the means for a decent subsistence. The denial of the basic rights of particular other species will not, however, assist the world's poor and oppressed to win their just struggles. Nor is it reasonable to ask that the members of these other species should wait until all humans have achieved their rights first. That suggestion itself assumes that beings belonging to other species are of lesser moral significance than human beings. Moreover, on present indications, the suggested delay might well be an extremely long one.

Another basis for opposition to our demand may arise from the fact that the great apes – especially chimpanzees – are considered to be extremely valuable laboratory tools. Of course, since the main object of research is to learn about human beings, the ideal subject of study would be the human being. Harmful research on non-consenting human beings is, however, rightly regarded as unethical. Because harmful research on non-consenting chimpanzees, gorillas or orang-utans is not seen in the same light, researchers are permitted to do things to these great apes that would be considered utterly abhorrent if done to human beings. Indeed, the value of the great apes as research tools lies precisely in the combination of two conflicting factors: on the one hand, the fact that, both physically and psychologically, they very closely resemble our own species; and on the other, the fact that they are denied the ethical and legal protection that we give to our own species.

Those who wish to defend the present routine treatment of the nonhuman great apes in laboratories and in other circumstances – disturbing details of which we present in this book – must now bear the burden of proof in refuting the case we make in these pages for including all great apes within the community of equals. If our arguments cannot be refuted, the way in which great apes other than humans are now treated will be shown to be an arbitrary and unjustifiable form of discrimination. For this, there will no longer be any excuse.

The resolution of a moral dispute is often just the beginning, not the end, of a social question. We know that, even if we can prove our view to be sound, we will still be far away from the moment when the dispersed members of the chimpanzee, gorilla and orang-utan species can be liberated and lead their different lives as equals in their own special territories in our countries, or free in the equatorial forests to which they once belonged. As normally happens when ethical progress runs its course, the obstacles will be many, and opposition from those whose interests are threatened will be strong. Is success possible? Unlike some oppressed groups that have achieved equality, chimpanzees, gorillas and orang-utans are unable to fight for themselves. Will we find the social forces prepared to fight on their behalf to bring about their inclusion within the community of equals? We believe that success is possible. While some oppressed humans have achieved victory through their own struggles, others have been as powerless as chimpanzees, gorillas and orang-utans are today. History shows us that there has always been, within our own species, that saving factor: a squad of determined people willing to overcome the selfishness of their own group in order to advance another's cause.

The Editors and Contributors

I

Encounters with Free-living Apes

1

Chimpanzees – Bridging the Gap

Jane Goodall

If there is a single person who has made people appreciate that chimpanzees are individuals with different personalities and complex social relationships, that person is Jane Goodall. Her book In the Shadow of Man, *based on her long familiarity with a community of chimpanzees in the Gombe region of Tanzania, was an international bestseller. More recently she has also published a detailed academic study of chimpanzees,* The Chimpanzees of Gombe, *as well as the popular* Through a Window. *In the following statement of her reasons for supporting the Declaration on Great Apes, Goodall draws on more than thirty years' experience of observing chimpanzees and thinking about the human–chimpanzee relationship.*

She was too tired after their long, hot journey to set to on the delicious food, as her daughters did. She had one paralysed arm, the aftermath of a bout of polio nine years ago, and walking was something of an effort. And so, for the moment, she was content to rest and watch as her two daughters ate. One was adult now, the other still caught in the contrariness of adolescence – grown up one moment, childish the next. Minutes passed. And then her eldest, the first pangs of her hunger assuaged, glanced at the old lady, gathered food for both of them and took it to share with her mother.

The leader of the patrol, hearing the sudden sound, stopped and stared ahead. The three following froze in their tracks, alert to the danger that threatened ever more sinister as they penetrated further into neighbouring territory. Then they relaxed: it was only a large bird that had landed in a tree ahead. The leader looked back, as though seeking approval for moving on again. Without a word the patrol moved on. Ten minutes later they reached a look-out place offering a view across enemy territory. Sitting close together, silent still, they searched for sign or sound that might indicate the presence of strangers. But all was peaceful. For a whole hour the four sat there, uttering no sound. And then, still maintaining silence, the leader rose, glanced at the others and moved on. One by one they followed him. Only the youngest, a youth still in his teens, stayed on for a few minutes by himself, reluctant, it seemed, to tear himself away from the prospect of violence. He was at that age when border skirmishes seemed exhilarating as well as challenging and dangerous. He couldn't help being fascinated, hoping for, yet fearing, a glimpse of the enemy. But clearly there would be no fighting that day and so he too followed his leader back to familiar haunts and safety.

We knew her as 'Auntie Gigi'. She had no children of her own, but two years ago she had more or less adopted two youngsters who had lost their own mother in an epidemic – pneumonia, probably. They were lucky, those two. Not that Gigi was all sweet and motherly, not at all. She was a tough old bird, somewhat mannish in many ways. But she made a perfect guardian for she stood no nonsense, not from anyone, and had high standing in her society. If anyone picked a quarrel with either of these two kids, he or she had Auntie Gigi to reckon with. Before Gigi came into the picture, one of the orphans, little Mel, had been cared for by Sam, a teenage youth. It was quite extraordinary – it wasn't even as though Sam was related to the sickly orphan. He had not even been close with Mel's mother during her life. Yet after she passed away, Sam and Mel became really close, like a loving father and child. Sam often shared his food with Mel, usually carried him when they went on long trips together, and even let the child sleep with him at night. And he did his best to keep him out of harm's way. Maybe it was because Sam's mother had got sick and died in that same epidemic. Of course, he'd not been spending much time with her then – he'd been out and about with the boys mostly. Even so, it is always a comfort if you can sneak off to Mum for a while when the going gets tough, and the big guys start picking on you. And suddenly, for Sam, his old mother wasn't there. Perhaps his closeness with that dependent little child helped to fill an empty place in his heart. Whatever the reason, Mel would almost certainly have died if Sam hadn't cared for him as he did. After a year Sam and Mel began spending less time together. And that was when Auntie Gigi took over.

Those anecdotes were recorded during our thirty-one years of observation of the chimpanzees of Gombe, in Tanzania. Yet the characters could easily be mistaken for humans. This is partly because chimpanzees do behave so much like us, and partly because I deliberately wrote as though I were describing humans, and used words like 'old lady', 'youth', and 'mannish'. And 'Sam' was really known as 'Spindle'.

One by one, over the years, many words once used to describe human behaviour have crept into scientific accounts of nonhuman animal behaviour. When, in the early 1960s, I brazenly used such words as 'childhood', 'adolescence', 'motivation', 'excitement', and 'mood' I was much criticised. Even worse was my crime of suggesting that chimpanzees had 'personalities'. I was ascribing human characteristics to nonhuman animals and was thus guilty of that worst of ethological sins – anthropomorphism. Certainly anthropomorphism can be misleading, but it so happens that chimpanzees, our closest living relatives in the animal kingdom, do show many human characteristics. Which, in view of the fact that our DNA differs from theirs by only just over 1 per cent, is hardly surprising.

Each chimpanzee has a unique personality and each has his or her own individual life history. We can speak of the history of a chimpanzee community, where major events – an epidemic, a kind of primitive 'war', a 'baby boom' – have marked the 'reigns' of the five top-ranking or alpha males we have known. And we find that individual chimpanzees can make a difference to the course of chimpanzee history, as is the case with humans. I wish there was space to describe here some of these characters and events, but the information, for those interested, can be found in my most recent book, *Through a Window*.[1]

Chimpanzees can live more than fifty years. Infants suckle and are carried by their mothers for five years. And then, even when the next infant is born, the elder child travels with his or her mother for another three or four years and continues to spend a good deal of time with her thereafter. The ties between family members are close, affectionate and supportive, and typically endure throughout life. Learning is important in the individual life cycle. Chimpanzees, like humans, can learn by observation and imitation, which means that if a new adaptive pattern is 'invented' by a particular individual, it can be passed on to the next generation. Thus we find that while the various chimpanzee groups that have been studied in different parts of Africa have many behaviours in common, they also have their own distinctive traditions. This is particularly well documented with respect to tool-using and tool-making behaviours. Chimpanzees use more objects as tools for a greater variety of purposes than any creature except ourselves and each population has its own tool-using cultures. For example, the Gombe chimpanzees use long, straight sticks from which the bark has been peeled to extract army ants from their nests; 100 miles to the south, in the Mahale Mountains, there are plenty of the same ants, but they are not eaten by the chimpanzees. The Mahale chimpanzees use small twigs to extract carpenter ants from their nests in tree branches; these ants, though present, are not eaten at Gombe. And no East African chimpanzee has been seen to open hard-shelled fruits with the hammer and anvil technique that is part of the culture of chimpanzee groups in West Africa.

The postures and gestures with which chimpanzees communicate – such as kissing, embracing, holding hands, patting one another on the back, swaggering, punching, hair-pulling, tickling – are not only uncannily like many of our own, but are used in similar contexts and clearly have similar meanings. Two friends may greet with an embrace and a fearful individual may be calmed by a touch, whether they be chimpanzees or humans. Chimpanzees are capable of sophisticated co-operation and complex social manipulation. Like us, they have a dark side to their nature: they can be brutal, they are aggressively territorial, sometimes they even engage in a primitive type of warfare. But they also show a variety of helping and care-giving behaviours and are capable of true altruism.

The structure of the chimpanzee brain and central nervous system is extraordinarily like ours. And this appears to have led to similar emotions and intellectual abilities in our two species. Of course, it is difficult to study emotion even when the subjects are human – I can only guess, when you *say* you are sad and *look* sad, that you *feel* rather as I do when I am sad. I cannot know. And when the subject is a member of another species, the task is that much harder. If we ascribe human emotion to nonhuman animals we are, of course, accused of anthropomorphism. But given the similarities in the anatomy and wiring of the chimpanzee and human brains, is it not logical to assume that there will be similarities also in the feelings, emotions and moods of the two species? Certainly all of us who have worked closely with chimpanzees over extended periods of time have no hesitation in asserting that chimpanzees, like humans, show emotions similar to – sometimes probably identical to – those which we label joy, sadness, fear, despair and so on.

Our own success as a species (if we measure success by the extent to which we have spread across the world and altered the environment to suit our immediate purposes) has been due entirely to the explosive development of the human brain. Our intellectual abilities are so much more sophisticated than those of even the most gifted chimpanzees that early attempts made by scientists to describe the similarity of mental process in humans and chimpanzees were largely met with ridicule or outrage. Gradually, however, evidence for sophisticated mental performances in the apes has become ever more convincing. There is proof that they can solve simple problems through process of reasoning and insight. They can plan for the immediate future. The language acquisition experiments have demonstrated that they have powers of generalisation, abstraction and concept-forming along with the ability to understand and use abstract symbols in communication. And they clearly have some kind of self-concept.

It is all a little humbling, for these cognitive abilities used to be considered unique to humans: we are not, after all, quite as different from the rest of the animal kingdom as we used to think. The line dividing 'man' from 'beast' has become increasingly blurred. The chimpanzees, and the other great apes, form a living bridge between 'us' and 'them', and this knowledge forces us to re-evaluate our relationship with the rest of the animal kingdom, particularly with the great apes. In what terms should we think of these beings, nonhuman yet possessing so very many human-like characteristics? How should we treat them?

Surely we should treat them with the same consideration and kindness as we show to other humans; and as we recognise human rights, so too should we recognise the rights of the great apes? Yes – but unfortunately huge segments of the human population are *not* treated with consideration and kindness, and our newspapers inform us daily of horrific violations of human rights in many countries around the world.

Still, things have got better in some Western-style democracies. During the past 100 years we have seen the abolition of enforced child and female labour, slavery, the exhibiting of deformed humans in circuses and fairs and many other such horrors. We no longer gather to gloat over suffering and death at public hangings. We have welfare states so that (theoretically) no one needs to starve or freeze to death and everyone can expect some help when they are sick or unemployed. Of course there are still a myriad of social injustices and abuses, but at least they are not publicly condoned by the government and, once public sympathy has been aroused, they are gradually addressed. We are trying, for example, to abolish the last traces of the old sadism in mental institutions.

Finally, there is a growing concern for the plight of nonhuman animals in our society. But those who are trying to raise levels of awareness regarding the abuse of companion animals, animals raised for food, zoo and circus performers, laboratory victims and so on, and lobbying for new and improved legislation to protect them, are constantly asked how they can devote time and energy, and divert public monies, to 'animals' when there is so much need among human beings. Indeed, in many parts of the world humans suffer mightily. We are anguished when we read of the millions of starving and homeless people, of police tortures, of children whose limbs are deliberately deformed so that they can make a living from begging, and those whose parents force them – even sell them – into lives of prostitution. We long for the day when conditions improve worldwide – we may work for that cause. But we should not delude ourselves into believing that, so long as there is human suffering, it is morally acceptable to turn a blind eye to nonhuman suffering. Who are we to say that the suffering of a human being is more terrible than the suffering of a nonhuman being, or that it matters more?

It is not so long ago, in historical perspective, that we abolished the slave trade. Slaves were taken from 'savage' tribes that inhabited remote corners of the earth. Probably it was not too difficult for slave traders and owners to distance themselves, psychologically, from these prisoners, so unlike any people their 'masters' had known before. And although they must have realised that their slaves were capable of feeling pain and suffering, why should that matter? Those strange, dark, heathen people were so *different* – not really like human beings at all. And so their anguish could be ignored. Today we know that the DNA of all ethnic groups of humans is virtually the same, that we are all – yellow, brown, black and white – brothers and sisters around the globe. From our superior knowledge we are appalled to think back to the intelligent and normally compassionate people who condoned slavery and all that it entailed. Fortunately, thanks to the perceptions, high moral principles and determination of a small band of people, human slaves were freed. And they were freed *not* because of sophisticated analysis of their DNA, but because they so obviously showed the same emotions, the same intellectual abilities, the same capacity for suffering and joy, as their white owners.

Now, for a moment, let us imagine beings who, although they differ genetically from *Homo sapiens* by about 1 per cent and lack speech, nevertheless behave similarly to ourselves, can feel pain, share our emotions and have sophisticated intellectual abilities. Would we, today, condone the use of those beings as slaves? Tolerate their capture and export from Africa? Laugh at degrading performances, taught through cruelty, shown on our television screens? Turn a blind eye to their imprisonment, in tiny barren cells, often in solitary confinement, even though they had committed no crimes? Buy products tested on them at the cost of their mental or physical torture?

Those beings exist and we *do* condone their abuse. They are called chimpanzees. They are imprisoned in zoos, sold to anyone who cares to buy them as 'pets', and dressed up and taught to smoke or ride bicycles for our entertainment. They are incarcerated and often tortured, psychologically and even physically, in medical laboratories in the name of science. And this is condoned by governments and by large numbers of the general public. There was a time when the victims in the labs would have been human; but thanks to a dedicated few who stood up to the establishment and who gradually informed the general public of the horrors being perpetrated behind closed doors, the insane and other unfortunates are now safe from the white-coated gods. The time has come when we must take the next step and protect our closest living relatives from exploitation. How can we do this?

If we could simply argue that it is morally wrong to abuse, physically or psychologically, any rational, thinking being with the capacity to suffer and feel pain, to know fear and despair, it would be easy – we have already demonstrated the existence of these abilities in chimpanzees and the other great apes. But this, it seems, is not enough. We come up, again and again, against that non-existent barrier that is, for so many, so real – the barrier between 'man' and 'beast'. It was erected in ignorance, as a result of the arrogant assumption, unfortunately shared by vast numbers of people, that humans are superior to nonhumans in every way. Even if nonhuman beings are rational and *can* suffer and feel pain and despair it does not matter how we treat them provided it is for the good of humanity – which apparently includes our own pleasure. They are not members of that exclusive club that opens its doors only to bona fide *Homo sapiens*.

This is why we find double standards in the legislation regarding medical research. Thus while it is illegal to perform medical experiments on a brain-dead human being who can neither speak nor feel, it is legally acceptable to perform them on an alert, feeling and highly intelligent chimpanzee. Conversely, while it is legally permitted to imprison an innocent chimpanzee, for life, in a steel-barred, barren laboratory cell measuring five foot by five foot by seven foot, a psychopathic mass murderer must be more spaciously confined. And these double standards exist only because the brain-dead patient and the mass murderer are *human*. They have souls and we cannot, of course, prove that chimpanzees have souls. The fact we cannot prove that *we* have souls, or that chimps do not, is, apparently, beside the point.

So how can we hope to procure improved legal standing for the great apes? By trying to prove that we are 'merely' apes, and that what goes for us, therefore, should go for them also? I see no point in altering our status as humans by constantly stressing that we differ from the apes *only* in that our brains are bigger and better. Admittedly at our worst we can outdo the Devil in wickedness, but at our best we are close to the angels: certain human lives and accomplishments vividly illustrate the human potential. As we plod from cradle to grave we need all the encouragement and inspiration we can get and it helps, sometimes, to know that wings and halos *can* be won. Nor do I think it useful to suggest reclassifying the great apes as *human*. Out task is hard enough without the waving of red flags.

Fortunately there are some heavy-duty people (like the editors of this book) out there fighting for the rights of the great apes, along with those fighting for the rights of humans. If only we could march under one banner, working for apes and humans alike, and with our combined intelligence and compassion – our humanity – strive to make ever more people understand. To understand that we should respect the individual ape just as we should respect the individual human; that we should recognise the right of each ape to live a life unmolested by humans, if necessary helped by humans, in the same way as we should recognise these rights for individual human beings; and that the same ethical and moral attitudes should apply to ape beings and human beings alike. Then, as the thesis of this book proposes, we shall be ready to welcome them, these ape beings, into a 'moral community' of which we humans are also a part.

Let me end with a combined message from two very special members of this moral community. The first is a chimpanzee being named Old Man. He was rescued from a lab when he was about twelve years old and went to Lion Country Safaris in Florida. There he was put, with three females, on an artificial island. All four had been abused. A young man, Marc Cusano, was employed to care for them. He was told not to get too close – the chimps hated people and were vicious. He should throw food to the island from his little boat. As the days went by Marc became increasingly fascinated by the human-like behaviour of the chimps. How could he care for them if he did not have some kind of relationship with them? He began going closer and closer. One day he held out a banana – Old Man took it from his hand. A few weeks later Marc dared step on to the island. And then, on a never to be forgotten occasion, Old Man allowed Marc to groom him. They had become friends. Some time later, as Marc was clearing the island, he slipped, fell and scared the infant who had been born to one of the females. The infant screamed, the mother, instinctively, leapt to defend her child and bit Marc's neck. The other two females quickly ran to help their friend; one bit his wrist, the other his leg. And then Old Man came charging towards the scene – and that, thought Marc, was the end. But Old Man pulled each of those females off Marc and hurled them away, then kept them at bay while Marc, badly wounded, dragged himself to safety.

'There's no doubt about it,' Marc told me later, 'Old Man saved my life'.

The second hero is a human being named Rick Swope. He visits the Detroit zoo once a year with his family. One day, as he watched the chimpanzees in their big new enclosure, a fight broke out between two adult males. Jojo, who had been at the zoo for years, was challenged by a younger and stronger newcomer – and Jojo lost. In his fear he fled into the moat: it was brand new and Jojo did not understand water. He had got over the barrier erected to prevent the chimpanzees from falling in – for they cannot swim – and the group of visitors and staff that happened to be there stood and watched in horror as Jojo began to drown. He went under once, twice, three times. And then Rick Swope could bear it no longer. He jumped in to try to save the chimp. He jumped in despite the onlookers yelling at him about the danger. He managed to get Jojo's dead weight over his shoulder, and then he crossed the barrier and pushed Jojo on to the bank of the island. He held him there (because the bank was too steep and when he let go Jojo slid back to the water) even when the other chimps charged towards him, screaming in excitement. He held him until Jojo raised his head, took a few staggering steps, and collapsed on more level ground.

The director of the institute called Rick. 'That was a brave thing you did. You must have known how dangerous it was. What made you do it?'

'Well, I looked into his eyes. And it was like looking into the eyes of a man. And the message was, "Won't *anybody* help me?"'

Old Man, a chimpanzee who had been abused by humans, reached across the supposed species barrier to help a human friend in need. Rick Swope risked his life to save a chimpanzee, a nonhuman being who sent a message that a human could understand. Now it is up to the rest of us to join in too.

Note

1. J. Goodall, *Through a Window: My Thirty Years with the Chimpanzees of Gombe* (Houghton Miflin, Boston, 1990).

2

Meeting a Gorilla

DOUGLAS ADAMS and MARK CARWARDINE

Douglas Adams's writing is a fertile source of new ways of looking at the world. His many extraordinary books include the enormously popular series that began with The Hitchhiker's Guide to the Galaxy, *which has also been both a radio and television series. The following passage comes from* Last Chance to See, *by Adams and Mark Carwardine, a zoologist. The book is based on the authors' travels around the world (with the assistance of the BBC) to see and describe the situation of endangered animals. Here Adams gives an account of meeting a free-living gorilla in Zaïre. The passage is reproduced by kind permission of the authors and the publishers, William Heinemann Ltd and Harmony Books, a division of Crown Publishers, Inc.*

The gorillas were not the animals we had come to Zaïre to look for. It is very hard, however, to come all the way to Zaïre and not go to see them. I was going to say that this is because they are our closest living relatives, but I'm not sure that's an appropriate reason. Generally, in my experience, when you visit a country in which you have any relatives living there's a tendency to want to lie low and hope they don't find out you're in town. At least with the gorillas you know that there's no danger of having to go out to dinner with them and catch up on several million years of family history, so you can visit them with impunity. They are, of course, only collateral relatives – nth cousins, n times removed. We are both descended from a common ancestor, who is, sadly, no longer with us, and who has, since Darwin's day, been the subject of endless speculation as to what manner of creature he/she was.

The section of the primate family of which we are members (rich, successful members of the family, the ones who made good and who should, by any standards, be looking after the other, less-well-off members of the family) are the great apes – we are great apes.

The other great apes are the gorillas (of which there are three subspecies: mountain, eastern lowland and western lowland), two species of chimpanzee and the orang-utans of Borneo and Sumatra. Of these, the most closely related are the gorillas, the chimps and us. We and the gorillas separated on the evolutionary tree more recently than our common ancestor separated from the orang-utans, so the gorillas are more closely related to us than they are to the orang-utans. We are very, very close relatives indeed – as close to each other as the Indian elephant and the African elephant, which also share a common, extinct ancestor.

The Virunga volcanoes, where the mountain gorillas live, straddle the border of Zaïre, Rwanda and Uganda. There are about 280 gorillas there, roughly two-thirds of which live in Zaïre, and the other third in Rwanda. I say roughly, because the gorillas are not yet sufficiently advanced in evolutionary terms to have discovered the benefits of passports, currency declaration forms and official bribery, and therefore tend to wander backwards and forwards across the border as and when their beastly, primitive whim takes them. A few stragglers even pop over into Uganda from time to time, but there are no gorillas actually living there as permanent residents because the Ugandan part of the Virungas only covers about twenty-five square kilometres, is unprotected and is full of people whom the gorillas, given the choice, would rather steer clear of.

Now, the business of tourism is obviously a vexed one. I had wanted to visit the gorillas for years, but had been deterred by the worry that tourism must be disturbing to the gorillas' habitat and way of life. There is also the risk of exposing the gorillas to disease to which they have no immunity. It is well known that the famous and extraordinary pioneer of gorilla conservation, Dian Fossey, was for most of her life passionately opposed to tourism and wished to keep the world away from the gorillas. However, she did, reluctantly, change her mind towards the end of her life, and the prevalent view now is that tourism, if it's carefully controlled and monitored, is the one thing that can guarantee the gorillas' future survival. The sad but unavoidable fact is that it comes down to simple economics. Without tourists it's only a question of which will happen first – either the gorillas' forest habitat will be entirely destroyed for crop farming and firewood, or the gorillas will be hunted to extinction by poachers. Put at its crudest, the gorillas are now worth more to the locals (and the government) alive than dead.

The restrictions, which are tightly enforced, are these. Each gorilla family can only be visited once a day, usually for about an hour, by a party of a maximum of six people, each of whom are paying US$100 for the privilege. And maybe they won't even get to see the gorillas.

We were lucky; we did. We were keeping very quiet and looking very carefully around us. There was nothing we could see near us, nothing in the trees above us, nothing peering furtively from the bushes. It was a moment or two before we saw anything at all, but then at last a slight movement caught our eyes. About thirty yards away down the track we were following, standing in plain view, was something so big that we hadn't even noticed it. It was a mountain gorilla, or perhaps I should say a gorilla mountain, standing propped up on its front knuckles so that it assumed the shape of a large and muscular sloping ridge tent.

You will have heard it said before that these creatures are awesome beasts, and I would like to add my own particular perception to this: these creatures are awesome beasts. It is hard to know how better to put it. A kind of humming mental paralysis grips you when you first encounter a creature such as this in the wild, and indeed there is no creature such as this. All sorts of wild and vertiginous feelings well up into your brain, that you seem to have no connection with and no name for, perhaps because it is thousands of millions of years since such feelings were last aroused.

The gorilla noticed us and stalked off into the undergrowth. We set off to follow him but he was in his own element and we were not. We were not even able to tell whereabouts in his own element he was and after a while we gave up and started to explore the area more generally again.

A little while later we came across a silverback lying on his side beneath a bush, with his long arm folded up over his head scratching his opposite ear while he watched a couple of leaves doing not very much. ('Silverback' simply means that the gorilla's back was silver, or grey-haired. Only the backs of males turn silver, and it happens after the male has reached maturity.) It was instantly clear what he was doing. He was mooching. It was quite obvious. Or rather, the temptation to find it quite obvious was absolutely overwhelming.

They look like humans, they move like humans, they hold things in their fingers like humans, the expressions which play across their faces and in their intensely human-looking eyes are expressions which we instinctively feel we recognise as human expressions. We look them in the face and we think, 'We know what they're like', but we don't. Or rather we actually block off any possible glimmering of understanding of what they may be like by making easy and tempting assumptions.

I crept closer to the silverback, slowly and quietly on my hands and knees, till I was about eighteen inches away from him. He glanced round at me unconcernedly, as if I was just someone who had walked into the room, and continued his contemplations. I guessed that the animal was probably about the same height as me – almost two metres – but I would think about twice as heavy. Mostly muscle, with soft grey-black skin hanging quite loosely on his front, covered in coarse black hair.

As I moved again, he shifted himself away from me, just about six inches, as if I had sat slightly too close to him on a sofa and he was grumpily making a bit more room. Then he lay on his front with his chin on his fist, idly scratching his cheek with his other hand. I sat as quiet and still as I could, despite discovering that I was being bitten to death by ants. He looked from one to another of us without any great concern, and then his attention dropped to his own hands as he idly scratched some flecks of dirt off one of his fingers with his thumb. I had the impression that we were of as much interest to him as a boring Sunday afternoon in front of the television. He yawned.

It's so bloody hard not to anthropomorphise. But these impressions keep on crowding in on you because they spark so much instant recognition, however illusory that recognition may be. It's the only way of conveying what it was *like*.

After a quiet interval had passed I carefully pulled the pink writing paper out of my bag and started to make the notes that I'm writing from at the moment. This seemed to interest him a little more. I suppose he had simply never seen pink writing paper before. His eyes followed as my hand squiggled across the paper and after a while he reached out and touched first the paper and then the top of my biro – not to take it away from me, or even to interrupt me, just to see what it was and what it felt like. I felt very moved by this, and had a foolish impulse to show him my camera as well.

He retreated a little and lay down again about four feet from me, with his fist once more propped under his chin. I loved the extraordinary thoughtfulness of his expression, and the way his lips were bunched together by the upward pressure of his fist. The most disconcerting intelligence seemed to be apparent from the sudden sidelong glances he would give me, prompted not by any particular move I had made but apparently by a thought that had struck him.

I began to feel how patronising it was of us to presume to judge their intelligence, as if ours was any kind of standard by which to measure. I tried to imagine instead how he saw us, but of course that's almost impossible to do, because the assumptions you end up making as you try to bridge the imaginative gap are, of course, your own, and the most misleading assumptions are the ones you don't even know you're making. I pictured him lying there easily in his own world, tolerating my presence in it, but, I think, possibly sending me signals to which I did not know how to respond. And then I pictured myself beside him, festooned with the apparatus of my intelligence – my Gore-Tex cagoule, my pen and paper, my autofocus matrix-metering Nikon F4, and my inability to comprehend any of the life we had left behind us in the forest. But somewhere in the genetic history that we each carry with us in every cell of our body was a deep connection with this creature, as inaccessible to us now as last year's dreams, but, like last year's dreams, always invisibly and unfathomably present.

It put me in mind of what I think must be a vague memory of a movie, in which a New Yorker, the son of East European immigrants, goes to find the village that his family originally came from. He is rich and successful and expects to be greeted with excitement, admiration and wonder.

Instead, he is not exactly rejected, not exactly dismissed, but is welcomed in ways which he is unable to understand. He is disturbed by their lack of reaction to his presence until he realises that their stillness in the face of him is not rejection, but merely a peace that he is welcome to join but not to disturb. The gifts he has brought with him from civilisation turn to dust in his hands as he realises that everything he has is merely the shadow cast by what he has lost.

I watched the gorilla's eyes again, wise and knowing eyes, and wondered about this business of trying to teach apes language. Our language. Why? There are many members of our own species who live in and with the forest and know it and understand it. We don't listen to them. What is there to suggest we would listen to anything an ape could tell us? Or that he would be able to tell us of his life in a language that hasn't been born of that life? I thought, maybe it is not that they have yet to gain a language, it is that we have lost one.

The silverback seemed at last to tire of our presence. He hauled himself to his feet and lumbered easily off into another part of his home.

3

Chimpanzees Are Always New to Me

TOSHISADA NISHIDA

Toshisada Nishida is professor of zoology at Kyoto University. Since 1965 he has been studying chimpanzees at Kasoje, at the foot of the Mahale Mountains, in Tanzania. His major interest has been in social organisation and relationships among chimpanzees. Here he briefly outlines why his long experience of watching chimpanzees has led him to support the Declaration on Great Apes.

There is no animal like a chimpanzee. There are many people who have studied chimpanzees in the wild for more than fifteen years. Once scientists are charmed by chimpanzees, then as long as they can continue their research, they never think of changing their object of study to another animal. On the other hand, probably no one has studied any other single species of animal for more than twenty years. Why this difference between the study of chimpanzees and other animals? The reason is simple and straightforward: chimpanzee behaviour is flexible, idiosyncratic and shows an extraordinarily varied repertoire.

Although I have spent several thousand hours with chimpanzees, I see forms of behaviour totally new to me whenever I go to Mahale. Even after twenty-five years of research, the chimpanzees of Mahale continue to be the source of astonishment, interest and pleasure for me. To take just one example, I shall describe some instances of 'water contact behaviour'.

Two years ago an adult male chimpanzee, Musa, was seen to 'wash' the skin of a colobus monkey in a stream. The previous evening there had been at least two colobus hunts, and in the morning the male, who had probably retained the skin overnight in his bed, began to scrape meat from it, and chew the coat. He brought the coat to the stream, where he repeatedly ducked it under the water and lifted it out. Next he sank it down to the bed of the stream for a while, and then shook it violently in the water. He even stood on two legs and stamped on the coat with one foot on the rock, as some people stamp on clothes to wash them. (However, he never wore the coat!) After the washing, he again nibbled meat and skin. Perhaps he had found the coat dirty and cleaned it; or perhaps he found it tough and was attempting to make the food more tender. Three males, both aged and young, watched Musa's new profession intently.

For decades, people have believed that chimpanzees in the wild are afraid of running water. It is true that they usually make every effort to avoid getting wet when they cross a stream. Once, however, a young female, Tula, stood on all fours near a pool in a running stream, and tried to bail the water out with one hand. Then she immersed herself in the water, splashed it and tried to uproot a submerged old buttress of a tree. She picked a large stone from the pool, carried it with both hands to a bigger, deeper pool and threw it into the water. She then stirred up the water with her hand.

One of the more startling displays performed by adult male chimpanzees of Mahale is to throw stones into the river. They aim at the surface of the water and more than 90 per cent of the stones actually hit the water. The splashing sound and the sight of a splash of water widely extended on the rocky surface appear to have a great intimidating effect on other chimpanzees. One can recognise idiosyncrasies in throwing technique. Some regularly use their right arms only, while others use either the right or the left. The present leading male uses both arms simultaneously to lift and throw the largest stone possible.

It is interesting to see the M Group chimpanzees crossing the largest river in our study area, which is about fifteen to twenty metres wide. They cross it by means of stepping stones. It is a risky endeavour, since the crossing provides raptors and lions with one of their best opportunities to catch and kill the most feeble members of the group. When the chimpanzees approach the river, those in front wait by the bank, mostly engaged in social grooming, until the others join them. As time passes, the scattered members of the group gradually gather together. Suddenly the leading male of the group appears to have decided to cross the river. Infants who usually walk on their own are carried on the belly of their mothers while crossing. A mother may have a newborn on her belly at the same time as she has her weaned infant on her back. As soon as the mother arrives at the opposite bank, the large infant climbs down to the ground.

My current research focuses on social relationships among adult males. The former leading or alpha male, Ntologi, retained his status for eleven years, before being ostracised by a coalition of adult males. The status of the current alpha male, Kalunde, is not yet stable. He has two male allies, one young (the fourth-ranking member of the group) and one old (the sixth-ranking member). As many as three adult males (the second-, third- and fifth-ranking members) appear to be his rivals for the leadership. The formation of cliques, and intervention against such formation, is an everyday occurrence that is too complex, subtle and flexible for me to summarise here. It might be said that male chimpanzees will tend to join forces against a strong rival. The story is, however, not so simple. Relationships among allies are often ambivalent, and they can be so intricate that it may require a decade of study to elucidate them.

The 'Declaration' at the beginning of this book proposes the inclusion of chimpanzees, gorillas and orang-utans in 'the moral community of equals'. Such an attempt is long overdue, given the similarity of humans and the other great apes, but it demands courage. Many people will protest against this proposal: some will say that human affairs are more important than anything else, while others will argue that the logical extension of including the great apes in the community of equals is the inclusion of all other life forms into that community.

I think we should extend the right of membership to other life forms where and when that becomes possible. But we can and should include the great apes in our moral community immediately, as a first step. Remember that for a long time people did not consider that even their neighbours belonged to their own kind. The concept of 'people' was applied only to members of one's own tribe. A British traveller who roamed around the Malay peninsula in the early 1900s believed that the naked indigenous hunter-gatherers he watched were not human beings, but a kind of anthropoid ape. He held to this belief, despite having watched these hunter-gatherers walking on two legs and using blowguns as a hunting weapon.

You might laugh at this British gentleman, saying that he lacked common sense. But can you really laugh? After another century has passed our descendants might laugh at those who hesitated to give basic moral rights to the great apes.

II

Conversations with Apes

4

Chimpanzees' Use of Sign Language

ROGER S. FOUTS and
DEBORAH H. FOUTS

Roger and Deborah Fouts founded the Chimpanzee and Human Communication Institute at Central Washington University, Ellensburg, Washington. Among the chimpanzees living at the Institute is Washoe, the first chimpanzee ever to communicate with human beings by means of a human language. Their personal account of their experiences with chimpanzees provides striking evidence of the remarkable intellectual abilities of these apes – and of the equally remarkable resistance of many human beings to the idea that other apes can be so like us.

Washoe was cross-fostered by humans.[1] She was raised as if she were a deaf human child and so acquired the signs of American Sign Language. Her surrogate human family had been the only people she had really known. She had met other humans who occasionally visited and often seen unfamiliar people over the garden fence or going by in cars on the busy residential street that ran next to her home. She never had a pet but she had seen dogs at a distance and did not appear to like them. While on car journeys she would hang out of the window and bang on the car door if she saw one. Dogs were obviously not part of 'our group'; they were different and therefore not to be trusted. Cats fared no better. The occasional cat that might dare to use her back garden as a shortcut was summarily chased out. Bugs were not favourites either. They were to be avoided or, if that was impossible, quickly flicked away. Washoe had accepted the notion of human superiority very readily – almost too readily. Being superior has a very heady quality about it.

When Washoe was five she left most of her human companions behind and moved to a primate institute in Oklahoma. The facility housed about twenty-five chimpanzees, and this was where Washoe was to meet her first chimpanzee: imagine never meeting a member of your own species until you were five. After a plane flight Washoe arrived in a sedated state at her new home. The director of the institute insisted that she be put in a cage in the main building housing the adult chimpanzees. Despite our protests he even took away her blanket, under the pretext that it was time she learned what it was to be a chimpanzee. The director was from the old, but still popular, school of captive treatment that explicitly held that humans had to dominate the animals they owned, and that the best way to do this was to arbitrarily mistreat them.

When Washoe awoke she was in a cage. After some argument, the director had grudgingly allowed one of us to stay, so she had at least one familiar friend with her when she woke. When she began to move, the chimpanzees in the adjoining cages began to bang and scream at her. After she regained her senses her human friend asked in sign language what the chimpanzees were. She called them 'BLACK CATS' and 'BLACK BUGS'. They were not like her and if she felt about them the way she felt about cats and bugs they were not well liked. Washoe had learned our arrogance too well.

However, it wasn't long before Washoe began to accept the other chimpanzees, and herself as one of them. Like Wendy from Peter Pan, she took on the role of mother to all the young ones as well as defender of the picked-upon underdog. She seemed to show genuine compassion for her newly discovered species. During her first year at the institute she was allowed to spend time on a small island that the young chimpanzees enjoyed. The island had been constructed with a steep red clay bank that went into a water moat; there was a three-foot-high electric fence on the island side of the moat. One day a new young chimp had arrived at the institute and the director put the chimp on the island. The chimp became quite distressed and tried to jump across the moat but landed in the middle of it. Washoe's reaction was interesting because this was a new chimp, one whom she hardly knew, but who was obviously in danger. The chimp went under the water and came up again. Washoe then jumped the electric fence and landed on a small grassy bank that extended about a foot from the fence. She held on to the bottom of one of the electric fence posts and stepped into the water, sliding down the steep submerged bank. She extended her hand to the drowning chimp and pulled her back to safety. Washoe had taken a great risk to save the stranger. It was truly a case of altruism on her part. If there was anyone she didn't like it was the arrogant humans who mistreated her friends. In the ten years she spent there she never gave up her self-worth, even though the director would occasionally try to intimidate her when her human friends were absent.

I have often wondered what it would be like to suddenly discover that you are not who you thought you were. Would we be like Washoe and accept it and show compassion and caring for our newly discovered conspecifics? Or would we maintain our earlier arrogance and continue to oppress and refuse to accept our own kind?

It could never happen to an individual human as it did to Washoe, but it has happened to all of us on another level: when Charles Darwin told us that those 'Black Bugs' were really our relatives. In reaction to this rude awakening some humans have clung to the vanity of human arrogance and continued to oppress and abuse their fellow animals. Others have discarded their false vanity and have attempted to remove the arrogance-induced ignorance by getting to know their newly discovered relatives. Some of us have even shown compassion and caring for them.

Human Arrogance

Why is human arrogance so pervasive and where does it come from? The answer to the first part of the question is easy. Arrogance is pervasive because it appeals to our vanity. We like it when we are told that we have high IQs, good looks and are extra special. What we seldom consider is that implicit in the statement that I have a high IQ is the suggestion that someone else must have a low IQ, and if I have good looks then surely someone else is quite ugly, and if I am extra special then most people must be quite ordinary. If this is so, then everything that is not me is sadly imperfect or downright defective. Once this attitude is established then you have a choice between advancing perfection or imperfection. There is no rational choice but to advance perfection. And what should you exploit in order to advance perfection? Why imperfection of course, those unfortunate individuals who are different from you. In this fashion you will become even more 'perfect' while at the same time removing some of life's imperfections.

Why do humans feel arrogant? It comes from our conception of animal nature. According to the seventeenth-century philosopher René Descartes, animals are unthinking, unfeeling machines, so different from us as to be uncomparable. How could we not help but become arrogant when Descartes justifies it? If this is true then it is important to ask where our conception of animal nature comes from. The answer is that our conception of animal nature does not come from the non-human animals themselves, but from our preconceived concepts of human nature.[2] We have not bothered to ask the animals what they are, but instead we tend to define them as not human. If humans have thought, animals don't; if humans have an imagination, animals have none; and so on. Many of us, in our reaction to the implications of Darwin's notion of continuity, try to maintain our false superiority by steadfastly clinging to our ignorance. We use the absence of evidence to claim evidence for absence with regard to sharing any traits that we think are important for our species' uniqueness.

Washoe, among other chimpanzees, has served notice on this studied ignorance spawned by human arrogance. The results of her accomplishments have put many academic feet in many academic mouths. Her accomplishments, along with those of her African cousins, have served as a small flame in the dark halls of human ignorance. It was only when a few humans were humble enough to ask the chimpanzee what their nature was that these discoveries were made. However, these discoveries have not always been well received because of the obvious conclusion that must be faced: namely, that we are no longer demiurges. Just as we are human beings, chimpanzees become chimpanzee beings and the importance shifts to the 'beingness'. Human is no longer a special classification but merely an adjective describing our animal nature.

Chimpanzee Mentality

The early days of Project Washoe set the stage for the fascinating discoveries to come later, and the Gardners set the example to follow: a combination of caring that took Washoe on her own terms and very rigorous experimental design.[3] Their use of double-blind testing procedures in subject-paced tests, and the careful diaries they kept of Washoe's daily activities, continue to be the highest standard for this area of research. The recent discoveries made with Washoe and her family today continue to add to this impressive record. We will present some of these more recent discoveries in order to shed a little light on our ignorance.

Cultural Transmission

When Washoe acquired her signs, some critics were quick to point out that her sign language was taught to her by humans and that she would not have acquired it without human intervention. They assumed that chimpanzees are incapable of passing information on from generation to generation, especially something as complex as language. In 1979, when Washoe was of child-bearing age, a study was done that would answer this premature criticism.

Washoe became pregnant and we designed a study to see whether she would pass her signs on to her offspring.[4] Judging from how readily captive chimpanzees imitate human skills, from the ability of wild chimpanzees to acquire tool-making skills from their friends and family, and especially from the fact that wild chimpanzees have demonstrated that they use gestural dialects which differ between chimpanzee communities,[5] it seemed likely that Washoe's infant would acquire signs from her.

Unfortunately, Washoe's own infant died, so a ten-month-old chimp from Yerkes Regional Primate Center was found to replace her dead infant and help ease her grieving. The infant's name is Loulis and Washoe readily adopted him. In order to control for the possibility that Loulis might acquire his signs from humans, we humans limited our signing in Loulis's presence to seven signs, otherwise we used vocal English to communicate with Washoe and Loulis, which she and he understood very well.

After the first eight days that Washoe and Loulis were together he began to imitate his first sign. Ten months is not an early age to learn signing; other signing chimpanzees have acquired their first signs in their fourth and fifth months of life. From our video recordings we found that Washoe was doing some very subtle teaching, in that she would initially orient towards Loulis, then sign COME, then approach him, and then retrieve him. She gradually faded this so that she stopped retrieving him and then she stopped approaching him and finally all she had to do was orient and sign. She also did some active teaching of signs. In one case she actually took his hand and moulded it into the sign for FOOD in an appropriate context. She was also observed to place a small toy chair in front of Loulis and then demonstrate the CHAIR/SIT sign. At fifteen months of age Loulis began to use his first two-sign combinations. What we found was that Loulis acquired his signs from Washoe and some of them she actively taught him. He used his signs to communicate with his fellow chimpanzees as well as with humans.

The tenacity of human ignorance was demonstrated after the early results of Washoe tutoring Loulis had been presented at the Psychonomics Society meetings in 1979, and even after several scientific articles had been published about Loulis's accomplishments (the first one in 1982). Even the attention of the popular media could not shake some scientists' hold on their ignorance. For example, as late as 1988 B. F. Skinner still felt able to publish the comment: 'No other species has developed the verbal environment we call a language. I doubt that the Gardners have ever seen one chimpanzee show another how to sign.'[6]

Chimpanzee Conversations

The next phase of our research looked at how Loulis used his signs with his mother and the other signing chimpanzees.[7] When we did this we had moved from Oklahoma, and Washoe and Loulis were joined by three chimpanzees whom the Gardners had cross-fostered on their second sign language project.[8] Moja joined Washoe and Loulis in 1979, and Dar and Tatu joined them in 1981. With five chimpanzees we were able to examine chimpanzee to chimpanzee sign language conversations. We found that Loulis gradually shifted his signing, as he grew older, from his adoptive mother Washoe to his new playmate Dar. This is typical of human children as well. What Loulis signed about with Dar was mainly play. They would request tickle and chase games from each other. However, whenever the game became too rough and one of them was hurt they would then sign to Washoe for comfort with HUG/LOVE signs and other solicitations of reassurance. When they fought we even observed Loulis apparently blaming Dar for the commotion. The two boys were screaming and fighting and when Washoe rushed in, as she usually did to stop the fights, Loulis signed to her 'GOOD GOOD ME' and then, screaming, pointed at Dar. Washoe would then discipline Dar. After several months of this Dar apparently caught on to the tactic and would throw himself on the floor when he saw Washoe enter the room and begin to cry and sign a frantic 'COME HUG' to her, whereupon she would comfort Dar. Then she would scold Loulis with a bipedal swagger toward him signing 'GO THERE' to him, pointing to the overhead exit tunnel for the room.

In one study we recorded over 5,200 instances of chimpanzee to chimpanzee signing.[9] This signing was analysed into different categories. The majority of signing by the chimpanzees occurred in the three categories of 'play', 'social interaction', and 'reassurance'; these accounted for over 88 per cent of the chimpanzee to chimpanzee conversations. The remaining 12 per cent was spread across the categories of 'feeding', 'grooming', 'signing to self', 'cleaning' and 'discipline'. An interesting aspect of these findings was that they indicated that the chimpanzees used their signs primarily for various types of social interaction. It also showed that food was not a major topic, since it accounted for only about 5 per cent of their conversations. Some critics who wished to discredit the chimpanzee language studies claimed that chimpanzee signing consisted solely of begging for food. (Although this was true of one study, in which the poor chimpanzee was deprived of his food and was required to sign in order to get it.[10])

The previous study was done using humans to observe the chimpanzees and record their behaviour in much the same fashion that Jane Goodall adopts in her observations of wild chimpanzees. However, in the tradition of careful experimental design that the Gardners began, we wished to control for any possible human intervention, so Debbi Fouts began a study that used remote video recording of the chimpanzees.[11] For this procedure, three or four cameras would be placed outside the enclosure in one of the chimpanzees' rooms and then connected by cables through the ceiling to monitors in a completely separate room. In this manner we were able to record the chimpanzees' signing with no human beings present. Debbi began in 1983, taking twenty-minute video samples three times a day for fifteen days. In these fifteen hours she found over 200 instances of chimpanzee to chimpanzee signing. Of course, in some twenty-minute samples there was absolutely none because the chimpanzees were napping during that particular random sample. However, in one of the twenty-minute samples there were twenty-nine chimpanzee to chimpanzee conversations. As with the live observation study, Debbi found that the chimpanzees talked mainly about their social interactions. She also found that when they talked about food it wasn't to obtain food. Instead, they merely talked about it, just as we might talk about some of our favourite foods without having to eat or even see them.

For three years Debbi continued to collect fifteen hours of video tape each year, to give a total of forty-five hours. She had several interesting records during this time. For example, Washoe did not seem to like the fact that everyone (all the humans at least) was required to leave the area during the taping. On one occasion after Debbi had positioned the cameras, shooed everyone out and gone back to the video monitoring room she saw on the screen that Washoe was approaching the cameras. Washoe then climbed up on the enclosure fence and looked directly into the camera and signed 'DEB DIRTY DEB'. Washoe uses the DIRTY sign to refer to faeces, soiled items or to humans or chimpanzees that she is displeased with.

Private Thoughts

In our live observation and subsequently in the remote video recording of the chimpanzees, we observed that they talked to themselves. This was not a new observation, since the Gardners had also noted that Washoe would do this when she was young. In fact, her private conversations with herself were truly private, even to the extent that if we tried to eavesdrop she would turn away; and if we continued to try to see what she was signing she would actually get up and move to a more secluded location. She would label pictures of things that she saw in magazines, or merely sign to herself. She would do this while alone in her bedroom, or to make sure she was not bothered sometimes she would take a magazine to the top of a thirty-foot willow tree and sign to herself up there. Later studies we did with the chimpanzees also found private signing. In the study with over 5,200 instances of chimpanzee conversations, 119 of these were private.[12]

When private signing occurs in humans it is considered to be overt thought – the person is thinking aloud. It is one of the few times that an observer can be privy to another person's private thoughts. Many philosophers and other intellectuals have claimed that thought is unique to humans and quite beyond the capacity of the other animals. This claim goes back to Aristotle, can be found in Aquinas and Descartes and has been defended by modern philosophers as well.[13] The research we are about to describe provides solid empirical evidence of nonhuman animal thought.

Debbi Fouts's forty-five hours of remote video tape had ninety instances of private signing on them. Mark Bodamer, one of our graduate students at the time, analysed these tapes for his master's thesis,[14] using research done with humans as a model. He found that, like humans, the chimpanzees used their private signing for a variety of functions. One major question that came out of Mark's thesis was whether the sample he analysed was biased against private signing because when Debbi recorded the tapes she would choose to record from cameras with two or more chimpanzees in the frame as opposed to single chimpanzees. She did this because she was focusing on chimpanzee to chimpanzee conversations. So after Mark's thesis we did another study, which collected twelve minutes of tape per day, five days a week, for a total of fifty-six weeks. We had fifty-six hours of new tape to analyse, but this time if those recording the session had a choice between recording from a camera with two chimpanzees in the frame as opposed to one that had a single chimpanzee, they were instructed to choose the latter. This procedure increased the number of private signing instances three-fold, to 368 for the 56 hours.

One of the more common categories of signing used by chimpanzees and humans is 'referential signing'. Examples are Washoe naming the picture in the magazine, or Dar signing DOG when he notices a dog running by outside his window. Basically, the individuals are simply commenting on things and events in their environment. They are doing it apparently just for the sake of it – they are not asking for it or begging for something. Some scientists have claimed that chimpanzees do not use referential communication but only sign for rewards. This arrogant position makes the chimpanzees seem more like unthinking machines than the active information-seeking beings they are in reality.

Another category demonstrates another behaviour that chimpanzees are not supposed to have. The claim is that chimpanzees only ask for things in their immediate environment and cannot ask for things not present. In other words, a kind of 'out of sight, out of mind' criticism. An utterance of private signing is categorised as 'informative signing' if the chimpanzees are referring to something not in their present environment. Again this is strong evidence that, just like us, the chimpanzees also think about things that are not present. The chimpanzees used this type of signing in 12 to 14 per cent of the instances in the two studies for a total of fifty-seven instances. This again demonstrates the rich mental life of the chimpanzee.

'Expressive signing' was popular with Washoe. This is a category of signing that is used more commonly by adult humans than by children. It occurs in humans when we become upset or excited by something. It might happen when you accidentally strike your thumb with a hammer, or if you notice a police car behind you just as you drive through a red light: even though you are alone you might say something very expressive. My favourite example of this for the chimpanzees was when Washoe was being recorded as she was lying on a bench looking at a magazine. Loulis came running into the room and into the camera frame from an overhead tunnel. He was running very fast and Washoe ignored him. As he ran under the bench where she was lying he reached up and stole her magazine and then ran immediately out of the room using the overhead tunnel again. By the time Washoe got to her feet, Loulis was gone. She then began to walk off and as she did she signed to herself 'DIRTY DIRTY'. As mentioned earlier, Washoe used the DIRTY sign as an insult.

Some of the other signs used by the chimpanzees were categorised as 'self-regulatory', 'regulatory', 'attentional', 'interactional', 'instrumental', 'describing own activity', 'question' and 'imaginary'. The 'imaginary' category was expanded upon by another of our students, Mary Lee Abshire, in a study on chimpanzee imagination that included imaginative private signing as well as other imaginative behaviours.[15]

Imaginative Chimpanzees

Imagination is another of those special mental behaviours that some people have considered unique to the human species. Some of our species' more impressive accomplishments have been attributed to imagination. For example, we might never have gone to the moon had we not imagined that it could be done. In the private signing study, imagination was defined as an utterance that is 'sung' or is word play, or represents a transformation of real objects or events, whether present or not: we found that 5 per cent of the utterances were imaginary. For example, rhythmic movements of signs or form alliteration of signs would be considered comparable to vocal singing. These were such events as Loulis playing with a block of wood by placing it on his head and then referring to it as a HAT. Another instance was when Moja produced an alliteration by 'rhyming' signs that all used the same initial hand configuration.

Mary Lee Abshire's thesis expanded on this to other behaviours, such as play. She found that chimpanzees, when playing, would treat toys as if they were alive. In other words, imagination involves attributing to situations or things certain properties that they do not actually possess. Using the remote video recording technique, she recorded Dar using a type of imagination referred to as 'animation' when he signed 'PEEKA-BOO' while playing with a teddy bear. Moja displayed a type of imagination referred to as 'substitution' when she began to treat a purse as if it were a shoe. Mary Lee was able to record six instances of imagination in the chimpanzees during fifteen hours of remote video recording. This is impressive when one considers that of the 5,200 observations of chimpanzee signing only about 2 per cent (119) were classified as private signing, and in the private signing studies only 4 to 5 per cent were imagination. In other words, imagination is a relatively rare behaviour compared with all the other things the chimpanzees do, just as it is with our species.

Timely Memories

Memory and a sense of time are two mental abilities that humans have thought absent in our fellow animals. Our favourite example of memory occurred with Washoe. About a year after we moved with Washoe to Oklahoma, the Gardners visited her. It was very hard on the Gardners to give Washoe up and send her away with us to Oklahoma, and it was perhaps because of the emotional pain associated with this that they did not visit Washoe again for another eleven years. By this time we had moved to our present home in Washington State and the chimpanzees from the Gardners' second project, Moja, Tatu and Dar, had joined Washoe and Loulis. Moja had not seen the Gardners for about three years and Dar and Tatu had not seen them for well over a year when they did visit. Loulis was the only chimpanzee who had never met them.

When the Gardners visited we did not tell the chimpanzees they were coming but kept it as a surprise. When they walked in the four chimpanzees who knew them did something very unusual. Normally if a stranger visits, the chimpanzee will begin to display and bang around in an apparent attempt to frighten the stranger away. When their familiar friends come in they usually greet us with pant hoots and Washoe and the others will often sign to us such things as COME HUG and want to touch us. However, when the chimpanzees saw the Gardners, except for Loulis they did neither of these things but sat down and stared at them as if they were dumbfounded. Loulis stood up and began to sway from side to side and bang the sides of the tunnel he was in, and his hair started to rise as well. When he started to do this Washoe and Dar, who were sitting on either side of him, both grabbed him. Dar covered his mouth with his hand and Washoe took his arm and shoulder and made him sit down and calm down. Loulis looked just as surprised by this as we were, because as far as we know the other chimpanzees had never treated him this way before.

The next surprise was when Washoe looked at the Gardners and signed their name signs. She had not seen them in eleven years, since she was seven years old, and she still remembered them and their name signs. Then Washoe signed 'COME MRS G' to Beatrice Gardner and led her into an adjoining room and began to play a game with her that she had not been observed to play since she was a five-year-old in Reno.

Another discovery we have made with Tatu combines memory with a sense of time. We have had only two examples of this and they were two years and nine months apart. The first one occurred during the Thanksgiving holidays in 1989. We make it a general rule here to celebrate all birthdays and holidays, since these seasonal events serve to break up the deadening routine that captive situations can have. We celebrate every Christmas by decking the halls with edible strings of dried fruits and treats in addition to the traditional tree, which is covered with edible strings of treats and edible ornaments as well. We always get the tree and decorate it on the weekend following the Thursday of Thanksgiving. The tree is placed outside the enclosure of one of the chimpanzee rooms, and as Christmas approaches the edible ornamentation grows and grows. Needless to say, the Christmas tree is a favourite topic of conversation with the chimpanzees, and they refer to it with a sign combination they devised – CANDY TREE. Then on Christmas Day the chimpanzees are given some of the ornaments to eat and, because there are so many, they continue to receive these as a daily treat until New Year's Day.

On the Friday following Thanksgiving in 1989 it began to snow outside, and it was on this occasion that Tatu asked the following question: 'CANDY TREE?' This impressed us a great deal because it could be interpreted that Tatu not only remembered the Christmas tree but also knew that this was the season for it, which is a temporal perception. However, we were also aware that this was but a single observation of this type of behaviour, and it was not until August 1991 that we made a second observation of a similar instance of behaviour.

As mentioned, we also celebrate all the birthdays each year. We have two birthdays right next to each other: Debbi Fouts's birthday is on the first of August and Dar's is on the second. This year we celebrated Debbi's birthday with treats and birthday songs as usual. Later that day, in the afternoon, Tatu asked 'DAR ICE CREAM?' Ice cream is often part of the birthday celebrations, and it appears that Tatu may have been aware of what came after Debbi's birthday.

Cain and Abel Revisited

Chimpanzees are our closest living relatives. In terms of biochemical similarities based on blood research[16] and genetic similarities[17] chimpanzees are actually closer to humans than they are to gorillas, even though all three primates are within 1 per cent of each other.

The similarities of the behaviour of the chimpanzee in the wild to human behaviour are just as striking as are the biochemical and genetic similarities. The work of Jane Goodall[18] and others has shown us that the behaviour of wild chimpanzees is not so different from that of non-technological groups of humans. Indeed, wild chimpanzees live in communities surrounded by traditional boundaries, they hunt, they care for their mothers (even to the extent of mourning themselves to death over their mother's death), they make tools and, perhaps most important of all, they can suffer from emotional as well as physical pain.

In addition to the marked similarities that their culture has to ours, there are also striking cognitive similarities. The Gardners found that chimpanzees have the capacity to acquire human sign language.[19] We have shown in this chapter that chimpanzees can pass this language on to the next generation, that they can use it spontaneously to converse with each other as well as with humans, that they can use their signs to think with, as evidenced by their private signing, that they have an imagination, that they have good memories and that they may even be able to perceive seasonal time.

This research with the chimpanzee, together with research with other great apes, demonstrates that the difference between apes and humans is one of degree and supports the Darwinian notion of continuity. This position runs counter to the more popular notion that humans are different in kind from other animals. Unfortunately, much of the biomedical research on chimpanzees assumes a kind of schizophrenic position: it justifies the use of chimpanzees as a medical model because of Darwinian continuity, and yet at the same time it claims moral immunity with regard to the physical and mental damage done to the chimpanzees on the basis that humans are different from other animals. As a result, the chimpanzees are treated as if they are unfeeling machines.

Over the past twenty-five years our own research has served to help transcend the popular idea that humans are different in kind from all other animals. We have demonstrated that chimpanzees are aware, that they feel, and that they have very rich mental lives. From now on, we humans have a responsibility to make sure that our relationship with our sibling species, the chimpanzee, as well as with other great apes, is not that of Cain and Abel, but instead follows the more humane tenet of 'love thy brother'.

Notes

1. R. A. Gardner and B. T. Gardner, 'A cross-fostering laboratory', in R. A. Gardner, B. T. Gardner and T. E. Van Cantfort (eds), *Teaching Sign*

Language to Chimpanzees (State University of New York Press, Albany, 1989), pp. 1–28.

2. H. Sarles, *After Metaphysics: Toward a Grammar of Interaction and Discourse* (Peter de Rieder Press, Lisse, The Netherlands, 1977), p. 27.

3. R. A. Gardner and B. T. Gardner, 'A test of communication', in R. A. Gardner, B. T. Gardner and T. E. Van Cantfort (eds), *Teaching Sign Language to Chimpanzees* (State University of New York Press, Albany, 1989), pp. 181–97.

4. R. S. Fouts, A. Hirsch and D. H. Fouts, 'Cultural transmission of a human language in a chimpanzee mother/infant relationship', in H. E. Fitzgerald, J. A. Mullins and P. Page (eds) *Psychobiological Perspectives: Child Nurturance Series*, vol. 3 (Plenum Press, New York, 1982), pp. 159–93; R. S. Fouts, D. H. Fouts and T. E. Van Cantfort, 'The infant Loulis learns signs from cross-fostered chimpanzees', in R. A. Gardner, B. T. Gardner and T. E. Van Cantfort (eds), *Teaching Sign Language to Chimpanzees* (State University of New York Press, Albany, 1989), pp. 280–92.

5. F. X. Plooij, 'Some basic traits of language in wild chimpanzees', in A. J. Lock (ed.), *Action, Gesture and Symbol: The Emergence of Language* (Academic Press, New York, 1978).

6. B. F. Skinner, 'Signs and countersigns', *Behavioral and Brain Sciences*, vol. 11, no. 3 (1988) pp. 466–7.

7. D. H. Fouts, 'Remote videotaping of a juvenile chimpanzee's sign language interactions within his social group' (unpublished master's thesis, Ellensburg, Washington, USA, Central Washington University, 1984); R. S. Fouts and D. H. Fouts, 'Loulis in conversation with cross-fostered chimpanzees', in R. A. Gardner, B. T. Gardner and T. E. Van Cantfort (eds) *Teaching Sign Language to Chimpanzees* (State University of New York Press, Albany, 1989), pp. 293–307.

8. Gardner and Gardner, 'A cross-fostering laboratory'.

9. R. S. Fouts, D. H. Fouts and D. Schoenfeld, 'Sign language conversational interactions between chimpanzees', *Sign Language Studies*, vol. 34 (1984) pp. 1–12.

10. H. Terrace, *Nim* (Knopf, New York, 1979).

11. D. H. Fouts, 'Remote videotaping of a juvenile chimpanzee's sign language interactions within his social group'; Fouts and Fouts, 'Loulis in conversation'.

12. Fouts *et al.*, 'Sign language conversational interactions'.

13. Aristotle, *De Anima II*, 3, 414 a (28)–415a (10); T. Aquinas, *Summa contra Gentiles*, Book III, Part II, ch. CXII; R. Descartes, *The Philosophical Works of Descartes*, translated by E. S. Haldane and G. R. T. Ross (Dover, New York, 1955); relevant extracts from these three philosophers are to be found in T. Regan and P. Singer (eds), *Animal Rights and Human Obligations*, 2nd edn (Prentice-Hall, Englewood Cliffs, NJ, 1989), pp. 4–19. For contemporary exponents of the view that animals cannot think, see R. G. Frey,

Interests and Rights: The Case Against Animals (Clarendon Press, Oxford, 1980) and Michael Leahy, *Against Liberation: Putting Animals in Perspective* (Routledge, London, 1991).

14. M. Bodamer, 'Chimpanzees signing to themselves' (unpublished master's thesis, Ellensburg, Washington, USA, Central Washington University, 1987); M. Bodamer, R. S. Fouts, D. H. Fouts and M. L. Abshire, 'Functional analysis of chimpanzee (*Pan troglodytes*) private signing', submitted to the *Journal of Comparative Psychology* (1993).

15. M. L. Abshire, 'Imagination in chimpanzees', *Friends of Washoe Newsletter*, vol. 9, no. 4 (1989) pp. 2–10.

16. M. E. King and A. C. Wilson, 'Evolution at two levels in humans and chimpanzees', *Science*, vol. 188 (1975) pp. 107–16.

17. R. Lewin, 'DNA reveals surprises in human family tree', *Science*, vol. 226 (1984) pp. 1179–83.

18. J. Goodall, *The Chimpanzees of Gombe: Patterns of Behavior* (The Belknap Press of Harvard University Press, Cambridge, MA, 1986).

19. Gardner and Gardner, 'A test of communication'.

5

Language and the Orang-utan: The Old 'Person' of the Forest

H. Lyn White Miles

Orang-utans are, as H. Lyn White Miles tells us in this chapter, the most mysterious of the great apes. We know less about orang-utans than we do about chimpanzees and gorillas. But Dr Miles knows at least one orang-utan, Chantek, very well because, as she now relates, she has spent several years teaching him sign language and conversing with him. The result is a portrait of an intelligent, inventive, reflective and sometimes judgemental being. Dr Miles teaches in the Department of Sociology/Anthropology at the University of Tennessee at Chattanooga. She is currently co-editing (with Robert Mitchell and N.S. Thomson) a book entitled Animals, Anecdotes and Anthropomorphism.

I still maintain, that his [the orang-utan] being possessed of the capacity of acquiring it [language], by having both the human intelligence and the organs of pronunciation, joined to the dispositions and affections of his mind, mild, gentle, and humane, is sufficient to denominate him a man.
Lord J. B. Monboddo, *Of the Origin and Progress of Language*, 1773[1]

If we base personhood on linguistic and mental ability, we should now ask, 'Are orang-utans or other creatures persons?' The issues this question raises are complex, but certainly arrogance and ignorance have played a role in our reluctance to recognise the intellectual capacity of our closest biological relatives – the nonhuman great apes.

We have set ourselves apart from other animals, because of the scope of our mental abilities and cultural achievements. Although there were religious perspectives that did not emphasise our estrangement from nature, such as the doctrine of St. Francis and forms of nature religions, the dominant Judeo-Christian tradition held that white 'man' was separate and was given dominion over the earth, including other races, women, children and animals. Western philosophy continued this imperious attitude with the views of Descartes, who proposed that animals were just like machines with no significant language, feelings or thoughts. Personhood was denied even to some human groups enslaved by Euro-American colonial institutions. Only a century or so ago, scholarly opinion held that the speech of savages was inferior to the languages of complex societies. While nineteenth-century anthropologists arrogantly concerned themselves with measurements of human skulls to determine racial superiority, there was not much sympathy for the notion that animals might also be persons.

Ignorance is almost always the basis for defining difference as 'other'. Since the West had no representatives of our closest relatives, the apes, we were ignorant of our primate heritage and the species that link us more closely with nature.[2] In cultures where humans could routinely observe apes, some very different world views emerged. For instance, the term 'orang-utan' in Malay has been variously translated as 'reasonable being of the woods'[3] or 'old person of the forest'.[4] The sense of 'orang' is one of intelligence, reverence and respect.[5] The Dyak of Malaysia have a myth that the orang-utan was an older form of person who wisely did not let humans know that they could speak for fear that they would be put to work.[6] Because the Dyak came into close contact with orang-utans, they understood that these creatures were a sort of 'person'.

In this century our ignorance is being transformed into a more mature understanding of our place in the cosmos. We have developed this awareness through investigating the nature of the universe, the evolution of life and our species; exploring the complexity of the human mind and body; moving towards human and animal rights; solving global problems of the environment, economy, politics and communication; and engaging in space exploration, with the implicit idea that we may find others like ourselves. Once Western science made the effort to look more closely at apes under natural conditions and in captivity, it became increasingly difficult to explain ape behaviour without making reference to human behaviour. Observers developed more complex methods to study primates and discovered a range of human-like behaviours, including family structures, tool-making, hunting, shelter construction, complex communication and deception.[7] Researchers were careful to avoid naive anthropomorphism, but there seemed to be strong similarities between ape and human culture. Frans de Waal felt compelled to describe the ape personalities and politics he observed in human terms, stating that they 'can only be portrayed accurately by using the same adjectives as we use to characterize our fellow human beings'.[8] The 'other' was beginning to resemble us more than we thought possible.

We are now at an ethical crossroads, propelled by remarkable discoveries about our close genetic relatedness to apes and the extent of ape intellectual and emotional abilities. We are wondering if apes have the mental capacity to have culture, learn to speak, reason, attribute knowledge to others, deceive, be self-aware, develop a sense of ethics – that is, are they persons? Darwin has shown that we are linked through common ancestry with our biological cousins, the apes. Thus, human characteristics and social structures have their roots deep in a primate heritage. There are, of course, important differences between humans and animals, but beliefs about the degree of human uniqueness have been challenged. One of these challenges has come from recent experiments to teach language to apes, the subject of my research for the last fifteen years. If an orang-utan can learn to use language, might it be a person? It is time to look deeply into the eyes of the 'other', and listen to what it has to say.

The Orang-utan

The orang-utan is the most mysterious of the great apes because relatively little is known about it. The pioneering primatologists Robert and Ada Yerkes described orang-utans as quiet, thoughtful and melancholy.[9] Prehistorically they had a much larger range than today, living in varied environments throughout Asia. Orang-utans are now found only on the islands of Borneo and Sumatra, where their habitat ranges from hilly and mountainous regions to swampy lowlands. Previously considered to be exclusively arboreal, or tree-dwelling, we now know that orang-utans travel long distances on the ground and visit caves. They are primarily fruit-eaters but have a wide diet, which may determine their large size and social organisation, which is more extensive than was once believed.[10]

There are significant differences between humans and great apes, but we share 98–99 per cent of our genetic make-up with them. If we were to strictly follow our own taxonomic system of classification, scientists would place the great apes in one genus with hominids (humans and near-human ancestors). But, an anthropocentric (human-centred) view prevails and humans are conveniently placed in a different genus from apes. Chimpanzees and gorillas are generally believed to be more closely related to humans than are orang-utans, primarily on the basis of genetic studies, although this has been questioned.[11] At odds with those studies, orang-utans have a surprising number of behavioural and biological similarities with humans, which has produced a puzzle. Of all the apes, orang-utans are most similar to humans in gestation period, brain hemispheric asymmetry, characteristics of dentition, sexual physiology, copulatory behaviour, hormonal levels, hair pattern, mammary gland placement and insightful style of cognition. Why would orangutans have these behavioural and morphological similarities to humans given their genetic distance?

One current explanation is that humans and African apes may have diverged from each other more recently, but that the chimpanzee and gorilla have evolved their own specialisations. Both the fossil data and comparisons of DNA and other biochemical measures suggest that the orang-utan is the most conservative, or primitive, of the great apes. They are most like the ancestral hominoid (ape-like primate) living about twelve million years ago that later gave rise to apes and humans.[12] Orang-utans have retained more of the characteristics of this hominoid than have the African apes. As a result, orang-utans have been labelled a 'living fossil', and thus are a kind of time traveller.[13]

Orang-utans have amazing abilities that need wider recognition within both the general population and the scientific community. Cognitive studies with orang-utans have shown that they are at least as intelligent as the African apes, and have revealed a human-like insightful thinking style characterised by longer attention spans and quiet deliberate action.[14] Susan Essock and Duane Rumbaugh commented: 'Chimpanzees are often reputed to be the "smartest" of the apes, and orang-utans have the reputation of being dull and sluggish. Such tags are unfortunate and contrary to the results of studies.'[15]

Orang-utans make shelters and other tools in their natural setting. In captivity, they learn to tie knots,[16] recognise themselves in mirrors,[17] use one tool to make another, and are the most skilled of the apes in manipulating objects.[18] They are the escape artists of zoos because of their ability to cleverly manipulate bolts and wires to get out of their enclosures, a trait with which I have become very familiar. In discussing these tendencies, Benjamin Beck has compared the probable use of a screwdriver by chimpanzees, gorillas and orang-utans. The gorilla would largely ignore it, the chimpanzee would try to use it in a number of ways other than as a screwdriver, and:

> The orang-utan would notice the tool at once but ignore it lest a keeper discover the oversight. If a keeper did notice, the ape would rush to the tool and surrender it only in trade for a quantity of preferred food. If the keeper did not notice, the ape would wait until night and then proceed to use the screwdriver to pick the locks or dismantle the cage and escape.[19]

Wright showed an orang-utan named Abang how to strike flakes from a piece of flint to make a knife, as our hominid ancestors did two million years ago. After Abang learned to make flakes, he opened a box containing food by cutting a string that held it closed.[20]

Finding that orang-utan and human brains are similar in areas specialised for language prompted scientists to speculate that orang-utans could possibly be taught to use gestural signs.[21] Since 1973, I have been doing just that, first with chimpanzees, and, more recently, with an orang-utan named Chantek. Now we do not have to wonder about what might be in the mind of apes, or what emotions they might feel. If we keep our expectations realistic and use human children as our model, we can just ask them. I have learned much about these creatures and, like my colleagues doing similar research, I have found myself unconsciously experiencing them as persons.

Chantek: An Orang-utan Who Uses Sign Language

The biochemical similarities between apes and humans seemed in conflict with our behavioural differences, until ape language experiments shifted scientific opinion and began to fill in the gap. Attempts to teach speech to orang-utans have not been very successful because apes lack the flexible right-angle bend to their vocal tract that is necessary to make the range of human speech sounds.[22] After researchers began to use American Sign Language for the deaf to communicate with chimpanzees and gorillas,[23] I began the first longitudinal study of the language ability of an orang-utan named Chantek, who was born at the Yerkes Primate Center in Atlanta, Georgia, USA.[24] There was criticism that symbol-using apes might just be imitating their human care-givers, but there is now growing agreement that orang-utans, gorillas and both chimpanzee species can develop language skills at the level of a two- to three-year-old human child.[25]

The goal of Project Chantek was to investigate the mind of an orang-utan through a developmental study of his cognitive and linguistic skills. It was a great ethical and emotional responsibility to engage an orang-utan in what anthropologists call 'enculturation', since I would not only be teaching a form of communication, I would be teaching aspects of the culture upon which that language was based. If my developmental project was successful, I would create a symbol-using creature which would be somewhere between an ape living under natural conditions and an adult human, which threatened to raise as many questions as I sought to answer.

Beginning at nine months of age, Chantek was raised at the University of Tennessee at Chattanooga by a small group of care-givers who communicated with him by using gestural signs based on the American Sign Language for the deaf. Chantek produced his first signs after one month and eventually learned to use approximately 150 different signs, forming a vocabulary similar to that of a very young child. Chantek learned names for people (LYN, JOHN), places (YARD, BROCK-HALL), things to eat (YOGURT, CHOCOLATE), actions (WORK, HUG), objects (SCREWDRIVER, MONEY), animals (DOG, APE), colours (RED, BLACK), pronouns (YOU, ME), location (UP, POINT), attributes (GOOD, HURT), and emphasis (MORE, TIME-TO-DO). We found that Chantek's signing was spontaneous and nonrepetitious. He did not merely imitate his care-givers as had been claimed for the sign language trained chimpanzee Nim; rather, Chantek actively used his signs to initiate communications and meet his needs.

Almost immediately, Chantek began to use his signs in combinations and modulated their meanings with slight changes in how he articulated and arranged his signs. He commented 'COKE DRINK' after drinking his coke, 'PULL BEARD' while pulling a care-giver's hair through a fence, 'TIME HUG' while locked in his cage as his care-giver looked at her watch, and 'RED BLACK POINT' for a group of coloured paint jars. At first he used signs to manipulate people and objects to meet his needs, rather than to refer to them. He knew the meaning of his signs the way a pet might associate a can of food or a word with feeding time. But, could he use these signs as symbols, that is, more abstractly to represent a person, thing, action or idea, even apart from its context or when it was not present?

One indication of the capacity to use symbolic language in both deaf and hearing human children is the ability to point, which some researchers argued that apes could not do spontaneously. Chantek began to point to objects when he was two years old, somewhat later than human children, as we might expect. First, he showed and gave us objects, and then he began pointing to where he wanted to be tickled and to where he wanted to be carried. Finally, he could answer questions like WHERE HAT?, WHICH DIFFERENT?, and WHAT WANT? by pointing to the correct object.

As Chantek's vocabulary increased, the ideas that he was expressing became more complex, such as when he signed 'BAD BIRD' at noisy birds giving alarm calls, and 'WHITE CHEESE FOOD-EAT' for cottage cheese. He understood that things had characteristics or attributes that could be described. He also created combinations of signs that we had never used before. In the way that a child learns language, Chantek began to over- or under-extend the meaning of his signs, which gave us insight into his emotions and how he was beginning to classify his world. For example, he used the sign 'DOG' for dogs, a picture of a dog in his viewmaster, orang-utans on television, barking noises on the radio, birds, horses, a tiger at the circus, a herd of cows, a picture of a cheetah, and a noisy helicopter that presumably sounded like it was barking. For Chantek, the sign BUG included crickets, cockroaches, a picture of a cockroach, beetles, slugs, small moths, spiders, worms, flies, a picture of a graph shaped like a butterfly, tiny brown pieces of cat food, and small bits of faeces. He signed 'BREAK' before he broke and shared pieces of crackers, and after he broke his toilet. He signed 'BAD' to himself before he grabbed a cat, when he bit into a radish, and for a dead bird.

We also discovered that Chantek could comprehend our spoken English (after the first couple of years we used speech as well as signing). One day when the radio was on, a children's story about a cat was being broadcast. When the narrator said 'cat' or made meow sounds, Chantek signed 'CAT'. We then verbally asked Chantek to sign a number of the words in his vocabulary, which he promptly did, showing that he had developed sign-speech correspondences without intentional training.

Another component of the capacity to use symbols is displacement: the ability to refer to things or events not present. It is an important indicator that symbols are also mental representations that can be held in the mind when the objects to which they refer are not present. This was an extremely important development in the evolution of human language because it freed individuals from the immediate environment and allowed our ancestors to talk about distant times and places. When he was two years old, Chantek began to sign for things that were not present. He frequently asked to go to places in his yard to look for animals, such as his pet squirrel and cat who served as playmates. He also made requests for 'ICE CREAM', signing 'CAR RIDE' and pulling us toward the parking lot for a trip to a local ice-cream shop.

We learned that an orang-utan can tell lies. Deception is an important indicator of language abilities since it requires a deliberate and intentional misrepresentation of reality. In order to deceive, you must be able to see events from the other person's perspective and negate his or her perception.[26] Chantek began to deceive from a relatively early age, and we caught him in lies about three times a week. He learned that he could sign DIRTY to get into the bathroom to play with the washing machine, dryer, soap, etc., instead of using the toilet. He also used his signs deceptively to gain social advantage in games, to divert attention in social interactions, and to avoid testing situations and coming home after walks on campus. On one occasion, Chantek stole food from my pocket while he simultaneously pulled my hand away in the opposite direction. On another occasion, he stole a pencil eraser, pretended to swallow it and 'supported' his case by opening his mouth and signing 'FOOD-EAT', as if to say that he had swallowed it. However, he really held the eraser in his cheek, and later it was found in his bedroom where he commonly hid objects.

We carried out tests of Chantek's mental ability using measures developed for human children. Chantek reached a mental age equivalent to that of a two- to three-year-old child, with some skills of even older children. On some tasks done readily by children, such as using one object to represent another and pretend play, Chantek performed as well as children, but less frequently. He engaged in chase games in which he would look over his shoulder as he darted about, although no one was chasing him. He also signed to his toys and offered them food and drink. Like children, Chantek showed evidence of animism, a tendency to endow objects and events with the attributes of living things. Although none of these symbolic play behaviours were as extensive as they would have been in a human child, the difference appears to be one of degree, not kind.

Chantek also experimented in play and problem-solving; for example, he tried vacuuming himself and investigated a number of clever ways to short out the electric fence that surrounded his yard. He learned how to use several tools, such as hammers, nails, and screwdrivers, and he was able to complete tasks using tools with up to twenty-two problem-solving steps. By the time he was two years old, he was imitating signs and actions. We would perform an action and ask him to copy it by signing 'DO SAME'. He would immediately imitate the behaviour, sometimes with novel twists, as when he winked by moving his eyelid up and down with his finger. Chantek also liked to use paints, and his own free-style drawings resembled those of three-year-old human children. He learned to copy horizontal lines, vertical lines and circles. By four and a half years of age, Chantek could identify himself in the mirror and use it to groom himself. He showed evidence of planning, creative simulation, and the use of objects in novel relations to one another to invent new meanings. For example, he simulated the context for food preparation by giving his care-giver two objects needed to prepare his milk formula and staring at the location of the remaining ingredient.

The above examples show evidence of intentionality, premeditation, taking the perspective of the other, displacement and symbolic use of language. These cognitive processes require that some form of mental image about the outcome of events be created. A further indication that Chantek had mental images is found in his ability to respond to his care-giver's request that he improve the articulation of a sign. When his articulation became careless, we would ask him to 'SIGN BETTER'. Looking closely at us, he would sign slowly and emphatically, taking one hand to put the other into the proper shape. Evidence for mental images also comes from Chantek's spontaneous execution of signs with his feet, which we did not teach him to do. Chantek even began to use objects in relation to each other to form signs. For example, he used the blades of scissors instead of his hands to make the sign for biting.

Chantek was extremely curious and inventive. When he wanted to know the name of something he offered his hands to be moulded into the shape of the proper sign. But language is a creative process, so we were pleased to see that Chantek began to invent his own signs. He invented: NO-TEETH (to show us that he would not use his teeth during rough play); EYE-DRINK (for contact lens solution used by his care-givers); DAVE-MISSING-FINGER (a name for a favourite university employee who had a hand injury); VIEWMASTER (a toy that displays small pictures); and BALLOON. Like our ancestors, Chantek had become a creator of language, the criterion that two hundred years earlier Lord Monboddo had said would define orang-utans as persons.

We had a close relationship with Chantek. He became extremely attached to his care-givers, and began to show empathy and jealousy toward us. He would quickly 'protect' us from an 'attacking' toy animal or other pretence. He clearly missed favourite care-givers and occasionally asked to see us. When he was eight years old, he became too large to live on campus, and he returned to the Yerkes Center in Atlanta, Georgia, to live. It was a difficult transition, and he missed his familiar companions and activities. One day he sat sadly, and signed 'POINT GIVE ANN' while gesturing toward the front door. He watched the door and the different cars and individuals that passed by – waiting for Ann. His loneliness was somewhat relieved when he was introduced to two female orang-utans at the Yerkes Center. Although he impregnated one of them, the offspring died shortly after birth. In the future, not only is it important that Chantek have an opportunity to continue to interact with other orang-utans, but it is also important that his enculturation not be forgotten. My goal is for our interaction to continue, and for Chantek to have an opportunity to use his signs not only with other humans, but with other orang-utans as well.

We have lived day to day with Chantek and have shared common experiences, as if he were a child. We have healed his hurts, comforted his fears of stray cats, played keep-away games, cracked nuts in the woods with stones, watched him sign to himself, felt fooled by his deceptions, and frustrated when he became bored with his tasks. We have dreamed about him, had conversations in our imagination with him and loved him. Through these rare events shared with another species, I have no doubt I was experiencing Chantek as a person.

Apes, Language and Personhood

How can Chantek's capabilities and our experiences with him help us to better understand how to define a 'person'? There are a number of abilities that have been suggested as necessary to personhood, and just as in the case of language, the definitions of personhood have become increasingly demanding as the boundaries of the abilities of *Homo sapiens* have been breached with the possibility that nonhumans may be persons. Descartes distinguished humans from animals on the basis of language and rational thought. Chantek's capabilities show that our species is not alone in having the ability to exhibit problem-solving intelligence, reason and mental representations. Furthermore, Chantek's language skills show that he was able to master a rudimentary communication system based on shared referential meanings that are conventionalised; abstracted from context; structurally interrelated; and expressed within a community of users to meet his needs, characterise his world, and influence the behaviour of others. This suggests that Chantek has met the Cartesian definition of person, at least at the level of a young human child.

In attempting to define personhood, psychologist Richard Passingham has added the additional requirements of invention, will, consciousness and conscience.[27] Michael Corballis[28] has suggested that will, consciousness and self-concept are perhaps difficult to define but, based on recent evidence of animal awareness, are unlikely to be unique to humans. Chantek's inventiveness is apparent as he creatively developed novel solutions to problems. Through creating his own signed words, he showed an even higher level of inventiveness, by generating symbols understood within the conventionalised framework of his language community. Chantek meets Passingham's requirement of exhibiting will through a pragmatic analysis of his communicative intentions. He clearly uses his signs to make requests, identify objects and comment on his world. His deceptions also reveal his will as he devises contrary means to achieve his goals. Passingham's third requirement, consciousness or self-awareness, has been investigated in animals and children through their ability to recognise themselves in a mirror. Although it became consistent at a later age, Chantek showed a pattern of self-recognition similar to human children.

Conscience is the final element in Passingham's definition, and it has also been stressed by philosopher Daniel Dennett. Dennett adds that self-reflection (the ability to self-consciously think about the self in relation to others), as well as an awareness that others have this ability, is also necessary to establish personhood.[29] Obviously, Chantek's childlike reflection and 'moral' behaviour are derivative and based on his enculturation by humans. But Chantek, like human children, did not reinvent nor genetically inherit his culture's ethical system. Over years of training children come to adopt their culture's pre-existing system and internalise it as their own.

Chantek's enculturation was based on the socially accepted behaviours for young children in educated American middle-class society. He was toilet trained, learned to eat with utensils, had to take turns in games and conversations, was distracted from masturbating in public, and not allowed to peek under the toilet doors at others using the facilities. We did not attempt to eradicate the 'orang-utan' in him by dressing him in nonfunctional human clothes or preventing him from exhibiting natural orang-utan behaviours, such as climbing trees or making his natural vocalisations. But he obviously developed a socially constructed self within the general boundaries of our culture. He knew his name and image, as well as the names and images of others. He used mirrors to groom himself and his care-givers. He could hold the existence of himself and others as a mental representation, engage in displaced reference and look for his companions, and occasionally ask for them by name when he was distressed. His deceptions, simulations and empathy showed that he could take the perspective of the other, based on a non-egocentric point of view, which he unquestionably had by four and a half years of age. This perspective-taking is the first step toward creating a concern for others and 'moral' behaviour.

Chantek's language ability also suggests that he internalised a minimal value system similar to that of a child. Chantek had several signs for emphasis and emotion, such as GOOD and BAD, which he used in appropriate contexts. The sign BAD is particularly interesting for he not only comprehended its meaning when he misbehaved, but he appropriated it to others and described their behaviour as BAD, as when he chastised noisy people, dogs and birds. He also labelled his own disapproved of behaviour as BAD, and even on occasion signed BAD to himself. Self-signing BAD is particularly interesting since it suggests that his purpose was to self-reflect; to have an internal dialogue about simple values. This reflection is nascent and immature, but Chantek has clearly internalised and exhibited some of the childlike 'morals' of his foster-culture.

With conscience and reflective self-awareness, we are dealing not only with the individual and his or her abilities, but also with the socially constructed self in conjunction with the beliefs and behaviours that form culture, including an ethical system. All human groups have ethical systems that play a role in their cultural adaptation, but the precise values selected and fostered are variable and dependent upon the infrastructural needs of the society. For instance, marrying several wives and ritually eating one's relatives or enemies are highly moral in one culture, but grounds for imprisonment or death in another. Conscience and morality are developed through enculturation, as one generation teaches values to another, otherwise we would not have the variability that we do in human cultures.

Apes, of course, adhere to their own patterns of behaviour within the constraints of their social order. These patterns are socially complex, rule-governed and based to a large extent on learning. Their acquired behaviour patterns are transmitted from generation to generation with variation from group to group in gestures, politics and social behaviours.[30] Although most learning is based on simple observation, there is some recent evidence for actual teaching by apes, as described in Chapter 4 and elsewhere, and for a degree of empathy or identification.[31] Because of this complexity we are increasingly inclined to describe the lifestyles of monkeys and apes under natural conditions as culture, or at least proto-culture. There is as yet no evidence that apes living freely have developed ethical systems based on extensive empathy. Nor is there yet evidence that apes have a theory of the mind, that is, an understanding that other individuals have beliefs and mental processes similar to their own. Ape 'cultures' tend to be simple, opportunistic, egocentric and pragmatic.[32] However, this is also the case for the behaviour of many humans, especially young children. When we decide if apes are persons, we should not require sentient beings to know of a human-based morality if they have not been exposed to one, because like human children, they may have the potential to develop one, however rudimentary.

Another approach to understanding personhood is to examine the definitions used by the law, particularly in terms of murder.[33] Chantek was raised in the states of Tennessee and Georgia: these states and many others define murder as the killing of 'reasonable creatures'. The Tennessee law formerly stated that murder is killing 'any reasonable creature in being', and currently simplifies this to killing 'another person'.[34] There have been several court cases in the United States that have addressed the issue of personhood.[35] One case in Mississippi discussed whether or not a slave was a person, and concluded that a slave was indeed a person, along with idiots and unchaste women.[36] It is interesting to note that these laws do not restrict murder to 'human beings' or 'Homo sapiens'. Thus, it could be argued that Chantek, or any other reasonable being, would be protected under this law.

In fact, a problem for those who require reflective self-awareness or full rational faculties (or the potential thereof) for personhood, or only the most extensive altruistic social behaviour, is that there are several categories of humans that do not meet this definition. Sociopaths, who can feel compassion for themselves but not for their victims, are self-reflective, but have not internalised a sense of cultural morality; they are familiar with their culture's morality, but their personal morality is purely egocentric. Severely mentally handicapped individuals and people who have extensive brain damage are not always self-reflective, yet we would consider them to be persons and protect them under the law.

We excuse children and mentally impaired people from adult responsibility, but we maintain that killing them (unless it is officially sanctioned by the state) is murder because of their 'potential' to have full human faculties, which may never be realised. Ethically speaking, enculturated apes are analogous to children. This analogy is particularly significant since the law protects children who show less linguistic and mental ability than Chantek.

Psychologists and philosophers search for definitions of 'personhood', but in fact, personhood is culturally defined by human groups as what feels like 'us', versus the 'other'. What is 'person' and what is the 'other' is relative, and varies in accordance with cultural definition. Because we have first-hand experience when we closely observe apes, we recognise those elements that suggest personhood. Chantek is no longer a free-living ape, nor is he a human being. Human language and enculturation have made him something in between – 'a reasonable being' – a person in both the Dyak sense and our own. During my years as an ape language researcher, I have seen many people gasp with amazement as they conversed with Chantek, subjectively experiencing him as a person. If it were possible for all humans to have this experience, this book might be unnecessary.

Acknowledgements – This research was supported by National Institutes of Health grant HD14918, National Science Foundation grant BNS 8022260, and grants from the UC Foundation. I would like to acknowledge the assistance of the Yerkes Regional Primate Research Center supported by National Institutes of Health grant 00165. I would also like to extend my gratitude to Stephen Harper in the preparation of this chapter, and my thanks to Robert Mitchell, Philip Lieberman, the members of Project Chantek and Chantek himself for their assistance in this research.

Notes

1. Lord J. B. Monboddo, *Of the Origin and Progress of Language*, Vol. 1. (1774), 2nd edn (reprinted, AMS Press, 1973), as cited in J. H. Schwartz, *The Red Ape* (Houghton Mifflin, Boston, 1987), p. 21.
2. E. Linden, *Apes, Men and Language* (Penguin Books, New York, 1981).
3. R. M. Yerkes and A. W. Yerkes, *The Great Apes: A Study of Anthropoid Life* (Yale University Press, New Haven, 1929), p. 38.
4. B. Galdikas-Brindamour, 'Orangutans, Indonesia's "People of the Forest" ', *National Geographic*, vol. 148, no. 4 (1975) pp. 444–73.
5. D. Freeman, *The Great Apes* (Putnam, New York, 1979).
6. T. L. Maple, *Orang-utan Behavior* (Van Nostrand Reinhold, New York, 1980), p. 213.

7. D. L. Cheney and R. M. Seyfarth, *How Monkeys See the World: Inside the Mind of Another Species* (University of Chicago Press, Chicago, 1990); J. Goodall, *The Chimpanzees of Gombe: Patterns of Behavior* (The Belknap Press of Harvard University Press, Cambridge, MA, 1986); H. L. Miles, 'How can I tell a lie?: Apes, language and the problem of deception', in R. W. Mitchell and N. S. Thompson (eds), *Deception: Perspectives on Human and Nonhuman Deceit* (State University of New York Press, Albany, 1986), pp. 245–66.

8. F. de Waal, *Chimpanzee Politics: Power and Sex Among Apes* (Harper & Row, New York, 1982), p. 54.

9. Yerkes and Yerkes, *The Great Apes.*

10. B. Galdikas, 'Orangutan diet, range, and activity at Tanjung Putting, Central Borneo', in *International Journal of Primatology*, vol. 9, no. 1 (1988), pp. 1–35.

11. H. Schwartz, *The Red Ape* (Houghton Mifflin, Boston, 1987).

12. D. Pilbeam, 'New hominoid skull material from the Miocene of Pakistan', *Nature*, vol. 295 (1982) pp. 232–4.

13. R. Lewin, 'Is the orangutan a living fossil?', *Science*, vol. 222 (1983) pp. 1222–3.

14. T. L. Maple, *Orang-utan Behavior.*

15. S. M. Essock and D. M. Rumbaugh, 'Development and measurement of the cognitive capabilities in captive nonhuman primates', in H. Markowitz, V. J. Stevens and L. P. Brett (eds), *Behavior of Captive Wild Animals* (Nelson-Hall, Chicago, 1978), pp. 161–208.

16. F. Jantschke, *Orang-utans in Zoologischen Garten* (R. Piper, Munchen, 1972).

17. G. G. Gallup, 'Towards an operational definition of self-awareness', in R. H. Tuttle (ed.), *Socioecology and Psychology of Primates* (Mouton, Paris, 1975).

18. J. Lethmate, 'Versuche zum "vorbedingten" Handeln mit einem jungen Orang-utan', *Primates*, vol. 18, no. 3 (1978) pp. 531–43; C. E. Parker, 'Responsiveness, manipulation and implementation behavior in chimpanzees, gorillas and orang-utans', in C. R. Carpenter (ed.), *Proceedings of the Second International Congress of Primatology*, vol. 1, *Behavior* (Karger Basel, New York, 1969), pp. 160–6.

19. B. B. Beck, *Animal Tool Behavior* (Garland Press, New York, 1980), pp. 68–9.

20. R. V. S. Wright, 'Imitative learning of a flaked stone technology – the case of an orangutan', *Mankind*, vol. 8, no. 4 (1972) pp. 296–306.

21. B. Galdikas, 'Living with the great orange apes', *National Geographic*, vol. 157, no. 6 (1980) pp. 830–53; M. LeMay and N. Geshwind, 'Hemispheric differences in the brains of great apes', *Brain Behavior and Evolution*, vol. 11 (1975) pp. 48–52.

22. W. Furness, 'Observations on the mentality of chimpanzees and orangutans', *Proceedings of the American Philosophical Society*, vol. 55, no. 3 (1916) pp. 281–90; K. Laidler, *The Talking Ape* (Stein and Day, New York, 1980); P. Lieberman, *The Biology and Evolution of Language* (Harvard University Press, Cambridge, MA, 1984).

23. R. A. Gardner and B. T. Gardner, 'Teaching sign language to a chimpanzee', *Science*, vol. 165 (1969) pp. 664–72; R. S. Fouts, 'Acquisition and testing of gestural signs in four young chimpanzees', *Science*, vol. 180 (1973) pp. 978–80; F. G. Patterson, 'Linguistic capabilities of a lowland gorilla', in F. C. C. Peng (ed.), *Sign Language and Language Acquisition in Man and Ape: New Dimensions in Comparative Pedolinguistics* (Westview Press, Boulder, CO, 1978); H. S. Terrace, L. Petitto, R. Sanders and T. Bever, 'Can an ape create a sentence?' *Science*, vol. 206 (1979) pp. 809–902.

24. H. L. Miles, 'Acquisition of gestural signs by an infant orangutan (*Pongo pygmaeus*)', *American Journal of Physical Anthropology*, vol. 52 (1980) pp. 256–7; H. L. Miles, 'Apes and language: The search for communicative competence', in J. deLuce and H. T. Wilder (eds), *Language in Primates: Perspectives and Implications* (Springer-Verlag, New York, 1983); H. L. Miles, 'The cognitive foundations for reference in a signing orangutan', in S. T. Parker and K. R. Gibson (eds), *'Language' and Intelligence in Monkeys and Apes: Comparative Developmental Perspectives* (Cambridge University Press, Cambridge, 1990), pp. 511–39.

25. H. L. Miles, 'The communicative competence of child and chimpanzee', in S. R. Harnard, H. D. Horst and J. Lancaster (eds), *Annals of the New York Academy of Sciences: Origins of Evolution of Language and Speech* (The New York Academy of Sciences, New York, 1976); Lieberman, *The Biology and Evolution of Language*.

26. R. Byrne and A. Whiten (eds), *Machiavellian Intelligence: Social Expertise and the Evolution of Intellect in Monkeys, Apes, and Humans* (Oxford University Press, Oxford, 1988); Mitchell and Thompson (eds), *Deception: Perspectives on Human and Nonhuman Deceit* (State University of New York Press, Albany, 1986).

27. R. Passingham, *The Human Primate* (W. H. Freeman and Company, Oxford, 1982), pp. 331–2.

28. M. Corballis, *The Lopsided Ape* (Oxford University Press, New York, 1991).

29. D. Dennett, 'Conditions of personhood', in *Brainstorms: Philosophical Essays on Mind and Psychology* (Bradford Books, Cambridge, MA, 1976/1978), pp. 267–85.

30. Goodall, *The Chimpanzees of Gombe*.

31. C. Boesch, 'Teaching among wild chimpanzees', *Animal Behaviour*, vol. 41 (1991) pp. 530–2; Goodall, *The Chimpanzees of Gombe*; D. J. Povinelli, K. E. Nelson and S. T. Boysen, 'Inferences about guessing and knowing by

chimpanzees (*Pan troglodytes*)', *Journal of Comparative Psychology*, vol. 104 (1990) pp. 203–10.

32. de Waal, *Chimpanzee Politics*.

33. R. B. Edwards and F. H. Marsh, 'Reasonableness, murder and modern science', *National Forum*, vol. 58 (1978) pp. 24–9.

34. *Tennessee Code Annotated*, 24:39–2401 (old), 39–13:201, and 39–13:202; see also *Tennessee Jurisprudence: An Encyclopedia of Tennessee Law*, 14:170 (The Michie Company, Charlottesville, VA, 1984).

35. State *versus* Jones, *Mississippi Reports*, vol. 1 (Courier and Journal Office: Natchez, Mississippi, 1821), p. 83.

36. Edwards and Marsh, 'Reasonableness, murder and modern science'.

6

The Case for the
Personhood of Gorillas

FRANCINE PATTERSON and
WENDY GORDON

*For twenty years Francine Patterson has been communicating
with Koko, a gorilla. Patterson began to study communication
with Koko by means of sign language in 1972. Basing her thesis
on this work, she received her doctorate in developmental psychology from Stanford University in 1979. Today she serves as
President of the Gorilla Foundation, which she and her associates
founded in 1976. This organisation, which serves as a trust on
behalf of Koko and two other gorillas, Michael and Ndume, is
currently working to establish a preserve in Hawaii where gorillas
will be able to live semi-free in a protected natural environment.
Wendy Gordon has worked at the Gorilla Foundation since 1990
as a research assistant, working regularly with both Koko and
Michael; before that she spent four years as a zoo volunteer,
educating the public about gorillas. This chapter describes some
of the interaction in the 'multi-species family' of gorillas and the
human beings who live and work with them.*

We present this individual for your consideration:
She communicates in sign language, using a vocabulary of over 1,000
words. She also understands spoken English, and often carries on
'bilingual' conversations, responding in sign to questions asked in
English. She is learning the letters of the alphabet, and can read some
printed words, including her own name. She has achieved scores
between 85 and 95 on the Stanford–Binet Intelligence Test.

She demonstrates a clear self-awareness by engaging in self-directed
behaviours in front of a mirror, such as making faces or examining her
teeth, and by her appropriate use of self-descriptive language. She lies to
avoid the consequences of her own misbehaviour, and anticipates
others' responses to her actions. She engages in imaginary play, both

alone and with others. She has produced paintings and drawings which are representational. She remembers and can talk about past events in her life. She understands and has used appropriately time-related words like 'before', 'after', 'later', and 'yesterday'.

She laughs at her own jokes and those of others. She cries when hurt or left alone, screams when frightened or angered. She talks about her feelings, using words like 'happy', 'sad', 'afraid', 'enjoy', 'eager', 'frustrate', 'mad' and, quite frequently, 'love'. She grieves for those she has lost – a favourite cat who has died, a friend who has gone away. She can talk about what happens when one dies, but she becomes fidgety and uncomfortable when asked to discuss her own death or the death of her companions. She displays a wonderful gentleness with kittens and other small animals. She has even expressed empathy for others seen only in pictures.

Does this individual have a claim to basic moral rights? It is hard to imagine any reasonable argument that would deny her these rights based on the description above. She is self-aware, intelligent, emotional, communicative, has memories and purposes of her own, and is certainly able to suffer deeply. There is no reason to change our assessment of her moral status if I add one more piece of information: namely that she is not a member of the human species. The person I have described – and she is nothing less than a person to those who are acquainted with her – is Koko, a twenty-year-old lowland gorilla.

For almost twenty years, Koko has been living and learning in a language environment that includes American Sign Language (ASL) and spoken English.[1] Koko combines her working vocabulary of over 500 signs into statements averaging three to six signs in length. Her emitted vocabulary – those signs she has used correctly on one or more occasions – is about 1,000. Her receptive vocabulary in English is several times that number of words.

Koko is not alone in her linguistic accomplishments. Her multi-species 'family' includes Michael, an eighteen-year-old male gorilla. Although he was not introduced to sign language until the age of three and a half, he has used over 400 different signs. Both gorillas initiate the majority of their conversations with humans and combine their vocabularies in creative and original sign utterances to describe their environment, feelings, desires and even what may be their past histories. They also sign to themselves and to each other, using human language to supplement their own natural communicative gestures and vocalisations.

Sign language has become such an integral part of their daily lives that Koko and Michael are more familiar with the language than are some of their human companions. Both gorillas have been known to sign slowly and repeat signs when conversing with a human who has limited signing

Table 1   Koko's performance on the Assessment of Children's
Language Comprehension (ACLC) test

Number of critical elements	Per cent correct				
	Chance	Sign + voice	Sign only	Voice only	Overall per cent
One – vocabulary (50 items)	20	72			
Two (e.g. HAPPY LADY)	20	70	50	50	56.7
Three (e.g. HAPPY LADY SLEEPING)	25	50	30	50	43.3
Four (e.g. HAPPY LITTLE GIRL JUMPING)	20	50	50	30	43.3
Overall per cent correct (two, three and four elements)		56.7	43.3	43.3	47.7

The results of χ^2 tests (1 df) indicate that Koko's performance on the ACLC in all modes and at all levels of difficulty was significantly better than chance, and that there was no significant difference in her comprehension whether the instructions were given in sign, English or sign plus English.

skills. They also attempt to teach as they have been taught. For example, one day Michael had been repeatedly signing 'CHASE' (hitting two fisted hands together) but was getting no response from his companion, who did not know this sign. He finally took her hands and hit them together and then gave her a push to get her moving. Similarly, Koko has often been observed moulding the hands of her dolls into signs.

Tests have shown that the gorillas understand spoken English as well as they understand sign. In one standardised test called the Assessment of Children's Language Comprehension, novel phrases corresponding to sets of pictures were given to the gorillas under conditions in which the tester did not know the correct answers. Koko's performance (see Table 1) was twice as good as might have been expected by chance, and there was no significant difference in her performance whether the instructions were given in sign only or in English only.[2]

Table 2   Koko's tested IQ 1975–6

Date	Test	CA*	MA+	IQ
Dec 1976	Khulman–Anderson	65	56	84.8
July 1976	Peabody Picture Vocabulary Test	60	49	81.6
Jan 1976	Stanford–Binet Intelligence Scale	54	46	85.2
Nov 1975	Wechsler Preschool and Primary Scale of Intelligence	51	37	71.0
July 1975	Stanford–Binet	48	44	91.7
Apr 1975	McCarthy Scales of Children's Abilities	45	32	73.0‡ (GCI)
Feb 1975	Stanford–Binet	43	37	86.0

★   Chronological age in months.
+   Mental age in months
‡   The McCarthy GCI stands for General Cognitive Index and is a scaled score, not a quotient.

Because the gorillas understand linguistic instructions and questions, we have been able to use standardised intelligence tests to further assess their abilities.[3] Koko's scores on different tests administered between 1972 and 1977 yielded an average IQ of 80.3 (see Table 2). More significant than the actual scores is the steady growth of Koko's mental age. This increase shows that she is capable of understanding a number of the principles that are the foundation of what we call abstract thought.

Many of those who would defend the traditional barrier between *Homo sapiens* and all other species cling to language as the primary difference between humans and other animals. As apes have threatened this last claim to human uniqueness, it has become more apparent that there is no clear agreement as to the definition of language. Many human beings – including all infants, severely mentally impaired people and some educationally deprived deaf adults of normal intelligence – fail to meet the criteria for 'having language' according to any definition. The ability to use language may not be a valid test for determining whether an individual has rights. But the existence of even basic language skills does provide further evidence of a consciousness which deserves consideration.

Conversations with gorillas resemble those with young children and in many cases need interpretation based on context and past use of the signs in question. Alternative interpretations of gorilla utterances are often possible. And even if the gorillas' use of signs does not meet a particular definition of language, studying that use can give us a unique perspective from which to understand more directly their physical and psychological requirements. By agreeing on a common vocabulary of signs we establish two-way communication between humans and gorillas. We can learn as much from what they say as we can by evaluating how they say it.

Some of what they tell us can be anticipated: 'What do gorillas like to do most?' 'GORILLA LOVE EAT GOOD'. Or, 'What makes you happy?' 'GORILLA TREE'. 'What makes you angry?' 'WORK'. 'What do gorillas do when it's dark?' 'GORILLA LISTEN [pause], SLEEP'. Some responses, on the other hand, are quite unexpected: 'How did you sleep last night?' (expecting 'FINE' 'BAD' or some related response). 'FLOOR BLANKET' (Koko sleeps on the floor with blankets). 'How do you like your blankets to feel?' 'HOT KOKO-LOVE'. 'What happened?' (after an earthquake). 'DARN DARN FLOOR BAD BITE. TROUBLE TROUBLE'.

Gorillas have suffered from a reputation for aloofness, low level of motivation and a contrary nature. Such gorilla stubbornness and negativism have been encountered and documented in our work with Koko and Michael, but certain findings indicate that this is evidence of intelligence and independence rather than of stupidity. And it is just this ornery independence that seems to spark episodes of humour and verbal playfulness. A characteristic incident involved Koko and assistant Barbara Hiller. Koko was nesting with a number of white towels and signed, 'THAT RED', indicating one of the towels. Barbara corrected Koko, telling her that it was white. Koko repeated her statement with additional emphasis, 'THAT RED'. Again Barbara stated that the towel was white. After several more exchanges, Koko picked up a piece of red lint, held it out to Barbara and, grinning, signed, 'THAT RED'.

Our approach has been to give Koko and Michael vocabulary instruction but no direct teaching of any other language skill. Most of the signs were learned either through the moulding of the gorillas' hands into signs or through imitation. But Koko and Michael have both created signs and used the language in diverse ways not explicitly taught. In a very real sense, the study has involved the mapping of skills, rather then the teaching of skills. This mapping is being done through observations in relatively unstructured and uncontrolled situations and through rigorous tests. The best possible linguistic and cognitive performances are likely to be given in the informal setting, with support coming from tests.

The gorillas have taken the basic building block of conversation (signs) and, on their own, added new meaning through modulation, a grammatical process similar to inflection in spoken language. A change in pitch or loudness of the voice, or the addition (or substitution) of sounds, can alter the meaning of a spoken word. In sign language this is accomplished through changes in motion, hand location, hand configuration, facial expression and body posture. The sign BAD, for instance, can be made to mean 'very bad' by enlarging the signing space, increasing the speed and tension of the hand, and exaggerating facial expression. Koko, like human signers, has exploited this feature of sign language to exaggerate a point, as when she signed THIRSTY from the top of her head to her stomach, instead of down her throat.

The gorillas have been observed to use these kinds of variations to mark relations of size (e.g. small versus large ALLIGATOR sign), number (BIRDS versus BIRD by repeating the sign), location (SCRATCH-ON-BACK), possession (KOKO'S-BABY signed simultaneously), manner, degree, intensity or emphasis (TICKLE signed with two hands), agent or object of an action (YOU-SIP signed by moving the signing hand toward the intended agent), negation (negating the ATTENTION sign by changing its location), to express questions (through eye contact and facial expression) and as a form of word play akin to wit or humour (simultaneously signing SAD FROWN when asked to 'smile' for the camera).

A conversation with Koko that involved this kind of creativity with the sign ROTTEN has been documented on film. Koko demonstrated the standard form of the sign in an exchange of insults after her companion called her a STINKER. Koko then inflected the sign by using two hands (perhaps meaning 'really rotten') and in the same sequence, brought the sign off her nose toward her companion, conveying the idea 'you're really rotten'. Koko's use of ROTTEN in this conversation also demonstrates her grasp of the connotation of a word rather than its denotation or concrete or specific meaning.

The meaning of the gorillas' signs are not necessarily identical to the most common meanings, and in some cases they are entirely different. To interpret the gorillas' conversations correctly, one must discover what the signs mean to them. Word-association games provide some clues. The gorillas are asked, 'What do you think of when I say _____?' Examples of stimulus-response pairs from sessions with Michael include: TEACHER-HAND, MICHAEL-FOOT, and similarly, AFRAID-HUG, DEAD-STINK and HUNGRY-EAT. The gorillas have also simply been asked to give definitions. Examples from data on Koko follow:

12 February 1984
Teacher: What's an insult?
Koko: THINK DEVIL DIRTY.

T: What's a stove?
K: COOK WITH
12 July 1984
T: What's an injury? [Voiced only.]
K: THERE BITE [to a cut on her hand].
13 July 1984
T: What is crazy?
K: TROUBLE SURPRISE.
8 February 1985
T: When do people say darn?
K: WORK OBNOXIOUS.
21 April 1983
T: What can you think of that's hard?
K: ROCK . . . WORK.
9 February 1984
T: What's a smart gorilla?
K: ME.

There are also words that Koko does not understand, and she sometimes corrects her companions when they apply them to her: on Christmas Eve in 1984 Koko picked up the telephone, listened to the dial tone, then signed 'RON' and handed the receiver to Barbara Hiller. When another companion commented, 'She's a goofball!' Koko responded, 'NO, GORILLA'. She has given similar responses when referred to as a 'juvenile' or a 'genius'.

Another way Koko and Michael have created novel meanings for basic vocabulary signs is through an unusual coining process in which they employ signs whose spoken equivalents match or approximate the sounds of English words for which no signs have been modelled. For example, Koko uses a modulated *knock* sign to mean 'obnoxious'. This indicates that she knows:

1. That the sign KNOCK is associated with the spoken word 'knock'.
2. That 'knock' sounds like the spoken word 'obnoxious'.
3. That the sign KNOCK can therefore be applied semantically to mean something or someone obnoxious.

Other examples include the substitution of the sign TICKLE for 'ticket', SKUNK for 'chunk', and LIP STINK for 'lipstick'. When Michael was asked to 'say bellybutton', he first signed 'BELLYBUTTON' (pointing to it), then signed 'BERRY BOTTOM'.

When signs have been repeatedly demonstrated that are difficult or impossible for Koko to form, her solution has often been to make substitutions based on the sound of the corresponding English word: KNEE for 'need', RED for 'thread', LEMON for 'eleven', and BIRD for 'word'.

The gorillas also communicate new meanings by making up their own entirely new signs. The intended meanings of some of the gorillas' invented nouns have been obvious ('nailfile', 'eyemakeup', 'barrette') because of their iconic form. The meanings of more abstract words such as verbs and prepositions ('above', 'below', 'take-off'), have to be worked out over time from records of the situations in which they occurred.

An analysis of the 876 signs emitted by Koko during the first ten years of the project[4] revealed that fifty-four signs, 6 per cent of her total emitted vocabulary, were her own inventions. Another 2 per cent (fifteen signs) were compounded by Koko from signs she was taught. Originally, only ten signs (1 per cent) were counted as natural gorilla gestures. New data from detailed observations of the gestures used by uninstructed gorillas indicates that these categories are fluid, and some of Koko's inventions are shared by other gorillas.

These invented signs indicate that the gorillas, like human children, take initiative with language by making up new words and by giving new meanings to old words. On the next level, there is evidence that Koko and Michael can generate novel names by combining two or more familiar words. For instance, Koko signed 'BOTTLE MATCH' to refer to a cigarette lighter, 'WHITE TIGER' for a zebra, and 'EYE HAT' for a mask. Michael has generated similar combinations, such as 'ORANGE FLOWER SAUCE' for nectarine yogurt and 'BEAN BALL' for peas. Other examples in the samples of the gorillas' signing are 'ELEPHANT BABY' for a Pinocchio doll and 'BOTTLE NECKLACE' for a six-pack soda can holder. Critics have commented that such phrases are merely the pairing of two separate aspects of what is present. Many of the above examples, however, cannot be explained in this way – when Koko signed 'BOTTLE MATCH', neither a bottle nor a match was present.

The gorillas have applied such new descriptive terms to themselves as well as to novel objects. When angered, Koko has labelled herself a 'RED MAD GORILLA'. Once, when she had been drinking water through a thick rubber straw from a pan on the floor after repeatedly asking her companion for drinks of juice which were not forthcoming, she referred to herself as a 'SAD ELEPHANT'.

Intrigued by examples of language use such as these, we went on to obtain more empirical evidence of the gorillas' metaphoric capacity. Koko was given a test devised by Howard Gardner[5] in which she was asked to assign various descriptive words to pairs of colours. The adjective pairs used were 'light-dark', 'happy-sad', 'loud-quiet', 'hard-soft', and 'warm-cold'. In videotaped sessions administered under conditions in which Koko could see only the stimulus and not the experimenter, she had no difficulty identifying literal dark versus light (two shades of green), red (versus blue) as warm, brown (versus blue-grey) as hard, violet-blue (versus yellow-orange) as sad, and lemon yellow (versus spring green) as loud. Koko indicated her answers either by pointing or by verbal descriptions (e.g. 'ORANGE THAT FINE', when asked which colour was happy). Ninety per cent of her responses were metaphoric matches as determined by Gardner and by three project research assistants who took the same test. Preschoolers in Gardner's study made only 57 per cent metaphoric matches; seven-year-olds, 82 per cent.[6]

Another creative aspect of the gorillas' language behaviour is humour. Humour, like metaphor, requires a capacity to depart from what is strictly correct, normal or expected. For example, when asked to demonstrate her invented sign for STETHOSCOPE for the camera, Koko did it on her eyes instead of on her ears. Asked to feed her chimp doll, she put the nipple to the doll's eye and signed 'EYE'. Appreciation of this kind of wit is sometimes dependent on recognising the sign behind the distortion. A sceptic might see this as a simple error, but in the case of signs that the gorillas themselves invent, such as STETHOSCOPE, this is not likely, and there are consistencies that run across the gorillas' humorous use of signs.

We have often noticed Koko giving an audible chuckling sound at the result of her own and her companions' discrepant statements or actions. She discovered that when she blew bugs on her companions, a predictable shrieking and jumping response could be elicited. Originally, she laughed at this outcome, but now she chuckles in anticipation of the prank as well. Accidents and unexpected actions by others can also cause Koko to laugh. Chuckles were evoked, for instance, by a research assistant accidentally sitting down on a sandwich and by another playfully pretending to feed sweets to a toy alligator. Developmental psychologists have found that the earliest form of humour in young children, incongruity-based humour, relies on similar principles of discrepancy applied to objects, actions and verbal statements.

Koko has also made verbal 'jokes'. On 30 October 1982, Barbara Hiller showed Koko a picture of a bird feeding her young.

K: THAT ME [to the adult bird].
B: Is that really you?
K: KOKO GOOD BIRD.
B: I thought you were a gorilla.
K: KOKO BIRD.
. . .
B: Can you fly?

K: GOOD. [GOOD can mean yes.]
B: Show me.
K: FAKE BIRD, CLOWN. [Koko laughs.]
B: You're teasing me. [Koko laughs.]
B: What are you really?
Koko laughs again, and after a minute signs
K: GORILLA KOKO.

In stark contrast to the gorillas' ability to express humour is their ability to communicate their thoughts and feelings about death. When Koko was seven, one of her teachers asked, 'When do gorillas die?' and she signed, 'TROUBLE, OLD.' The teacher also asked, 'Where do gorillas go when they die?' and Koko replied, 'COMFORTABLE HOLE BYE.' When asked 'How do gorillas feel when they die – happy, sad, afraid?' she signed, 'SLEEP'. Koko's reference to holes in the context of death has been consistent and is puzzling since no one has ever talked to her about burial, nor demonstrated the activity. That there may be an instinctive basis for this is indicated by an observation at the Woodland Park Zoo in Seattle, Washington. The gorillas there came upon a dead crow in their new outdoor enclosure, and one dug a hole, flicked the crow in, and covered it with dirt.[7]

In December of 1984 a tragic accident indicated the extent to which gorillas may grieve over the death of their loved ones. Koko's favourite kitten, All Ball, slipped out of the door and was killed by a speeding car. Koko cried shortly after she was told of his death. Three days later, when asked, 'Do you want to talk about your kitty?' Koko signed, 'CRY'. 'What happened to your kitty?' Koko answered, 'SLEEP CAT'. When she saw a picture of a cat who looked very much like All Ball, Koko pointed to the picture and signed, 'CRY, SAD, FROWN'. Her grief was not soon forgotten.

17 March 1985, with Francine Patterson
F: How did you feel when you lost Ball?
K: WANT.
F: How did you feel when you lost him?
K: OPEN TROUBLE VISIT SORRY.
F: When he died, remember when Ball died, how did you feel?
K: RED RED RED BAD SORRY KOKO-LOVE GOOD.

Arthur Caplan argues that animal interests and human interests should not be counted equally, claiming that nonhuman animals lack certain traits that make a moral difference. He uses the following example to illustrate his point:

If you kill the baby of a baboon the mother may spend many weeks looking for her baby. This behaviour soon passes and the baboon will go on to resume her normal life. But if you kill the baby of a human being the mother will spend the rest of her life grieving over the loss of her baby. Hardly a day will go by when the mother does not think about and grieve over the loss of her baby.[8]

But in this example the comparison is between outward behaviour in the case of the baboon mother, and a private mental state in the case of the human mother. In most such cases, the human mother also resumes her normal life: returning to her workplace, caring for her other children, going about her daily activities as before. Her grief is not necessarily apparent to the casual observer. Because the baboon mother cannot (or chooses not to) communicate *to us* her internal feelings about the death of her baby, it is assumed that it does not matter to her. While we cannot make any claims here about the emotional life of baboons, we have considerable evidence that Koko continues to mourn the loss of her adopted 'baby', All Ball, even years after his death.

19 March 1990
Koko comes across a picture of herself and All Ball in a photo album.
K: THAT BAD FROWN SORRY [emphatic] UNATTENTION.

Through conversations such as these the gorillas show not only that they are capable of experiencing emotions, but that they are aware of their emotions and can use language to describe them. Koko, at age six, was given a test that parallels a study with human children five to thirteen years old by Wolman, Lewis and King.[9] Koko was asked a series of questions with these frames:

(1) *Do you ever feel _____?*
(2) *When do you feel _____?*

The target feeling states were anger, fear, happiness, sadness, hunger, thirst, sleepiness and nervousness. Like the younger human subjects, Koko most frequently reported external events as conditions of emotional arousal; for example, when asked, 'When do you feel hungry?' she answered, 'FEEL TIME'. A possible explanation of this reply is that when it is time (to eat), she feels hungry. Koko regularly uses an emphatic TIME sign to tell her companion to bring out the next scheduled meal. Her replies to questions about anger seem to be related to events of the months preceding the test. Her responses to 'When do you feel mad?' included 'KOKO LOVE MARJIE BYE' and 'KOKO MAD GIRL'. At the time this test was given Koko had been having a difficult time with a new assistant named Marjie.

Koko has displayed a capacity for empathy in her comments about the emotional states of others:

24 September 1977
Koko is shown a picture of the albino gorilla Snowflake struggling against being bathed. Koko signs 'ME CRY THERE', indicating the picture.
3 November 1977, with companion Cindy Duggan
Koko looks at a picture of a horse with a bit in his mouth.
K: HORSE SAD.
CD: Why?
K: TEETH.
27 December 1977
Michael has been crying because he wants to be let out of his room. Koko, in the next room, is asked how Michael feels.
K: FEEL SORRY OUT.
7 April 1986
Mitzi Phillips tells Koko about a problem that is making her feel sad.
MP: What could I do to feel better?
K: CLOSE DRAPES . . . TUG-OF-WAR.
As Mitzi writes in the diary, Koko quietly comes up to her.
K: SAD? [Making the sign a question by raising her eyebrows and leaning forward, a standard ASL question form.]
MP: I feel better now.
Koko smiles.

The gorillas have also been asked to represent feeling states such as love, hate, happiness and anger with paints on canvas. Given free choice of ten or more colours, the gorillas produced works of contrasting colour and form. Asking them to paint emotions seemed a reasonable request because they had earlier demonstrated some primitive representational ability in their drawings and paintings done from models or from memory. Both Koko and Michael titled these works appropriately. One example of Michael's representational art is a work he called 'APPLE CHASE', for which he used our black and white dog named Apple as a model. The black and white painting bears a resemblance to Apple's head. (It is interesting to note that Michael and Apple have a special relationship. They frequently play 'chase' together, and Michael often initiates the game by signing CHASE to Apple.)

Michael frequently expresses himself creatively through sound play.
He uses various objects and parts of his body to produce a wide variety
of sounds and intricate rhythms. In creating his 'sound tools' he
experiments with different materials in his environment. In addition to
rhythmic drumming and tapping, for example, he sometimes strums a
rope or fabric strip held taut between his feet and his mouth. He made a
rattle by filling a PVC pipe end with hard nutshells and shaking it
vigorously with his hand covering the open end. Then he filled his
mouth with the nutshells and shook them around, making a contrasting
'wet' rattling sound.

Koko regularly expresses her creativity through fantasy play, alone or
with her companions. Often this play involves her plastic reptile toys
and centres on their tendency to 'bite'.

13 October 1988, with Mitzi Phillips
Koko is lying down with one of her toy alligators. She looks at it
and signs 'TEETH'. She examines its mouth. She kisses it, puts two
alligators together as if to make them kiss each other, then gives
them a three-way kiss. She puts her hand into the toy's mouth,
then pulls it out and shakes her hand.
MP: Oh, did it bite you?
K: BITE.
MP: Oh, no! Does it hurt?
Koko kisses her finger.
MP: May I see that bad alligator?
Koko gives it to Mitzi. Mitzi 'asks' the alligator why it bit Koko
and pretends to listen to its answer, then hands it back to Koko.
Koko kisses the toy again and again.
K: ALLIGATOR. GORILLA. BITE. GORILLA NUT NUT NUT. STOMACH
TOILET.

They are intelligent and emotional, they express themselves creatively
through language, art, music and fantasy play; but are gorillas self-
aware? Once considered unique to human beings, self-awareness is an
elusive concept. Its many definitions are both varied and vague,
although almost everyone has some notion of what it means. Through
their signing, Koko and Michael have shown a number of generally
accepted cognitive correlates of self-awareness, including the use of
personal pronouns, references to their own internal and emotional
states, humour, deception and embarrassment.

While self-awareness is probably best determined through the use of
language, self-recognition in mirrors is an accepted indicator of self-
awareness in human infants and other nonverbal individuals. In formal
mirror-marking tests, the subjects are first exposed to a mirror and

observed for any self-directed behaviour. Then their appearance is altered in such a way that they can only detect the change with the aid of a mirror. Nonhuman primates undergoing these tests are normally anaesthetised and marked on the face with red dye. Human children are marked surreptitiously with a spot of rouge while they are distracted. Once marked, subjects are again exposed to a mirror. Touching the mark while looking in the mirror is considered confirming evidence of self-recognition. Chimpanzees, orang-utans and humans have demonstrated a capacity for self-recognition in mirror tests, but the six gorillas previously tested failed to do so. It was concluded that gorillas lacked the cognitive capacity for self-awareness, in spite of informal reports to the contrary.

We gave a comparable mirror test to Koko[10] in which she demonstrated for the first time that gorillas, too, are capable of mirror self-recognition. For Koko, we used a modified procedure so that she would not have to be anaesthetised. During a series of ten-minute sessions videotaped over a three-day period, Koko's brow was wiped with a warm, damp, pink washcloth. During one of these sessions, the washcloth had been dipped in clown paint of the same pink colour. In the sessions in which she was unmarked, Koko touched the target area an average of only one time per session. During the fifth session when her brow was marked, she touched the target area forty-seven times, only after viewing it in the mirror. As she attempted to remove the paint, she also spent the most time viewing her reflection during the session in which she was marked. It is evident that Koko recognised the altered image as her own.

Koko had previously passed an informal 'mark test' when she attempted to rub away a dark spot of pigment on her upper gum, a spot that she had precisely located by looking into her mouth with a mirror. Captured on videotape, this spontaneous experiment of nature eliminated any possibility that Koko sensed the presence of the mark before noticing it in the mirror.

Why did the other gorillas fail to pass the mirror test? There are a number of possible explanations, including their age, rearing histories and social situations, their individual sensitivity to anaesthesia, or lack of motivation. There may also have been methodological problems, as at least two of the subjects touched the mark before exposure to the mirror. However, a more likely explanation is that the gorillas were inhibited by the presence of unfamiliar observers. Primatologists who have worked closely with gorillas have long been aware that the presence of strangers can profoundly affect gorilla behaviour, and this has been our experience as well. In certain situations Koko and Michael show a sensitivity to being watched even by familiar companions.

Ironically, it may have been the gorillas' very capacity for self-consciousness that prevented them from exhibiting behaviours indicative of self-recognition in the test situation.

Mirrors have been part of Koko's environment from her infancy. At the age of about three and a half, Koko began to exhibit self-directed behaviours in front of a mirror. She would groom her face and underarms, pick at her teeth and examine her tongue while studying her reflection. She would also comb her hair, make faces and adorn herself with hats, wigs and make-up. Michael has exhibited similar behaviours, which have been documented on videotape, even though his exposure to mirrors has been much more limited. These mirror-guided behaviours are normally exhibited by human children by the time they are two years old. Before that age children respond to their mirror image as they would to another child.

We took advantage of Koko's linguistic abilities to cross-validate the evidence of self-awareness provided by her response to mirrors. Two 'Who are you?' questions were asked during each of the first four test sessions while Koko was away from the mirror. After the fifth (marking) session Koko was asked this question once more while away from the mirror, and also 'Who is that?' of her mirror reflection. Although correct answers to these questions can simply be learned responses, the data argue otherwise in Koko's case. Her responses (listed in Table 3) were all different, multi-sign utterances, but had one consistent theme. Each reply to the questions contained one or more of the three signs: 'GORILLA', 'ME', 'KOKO'. Koko does not use these three words in response to all 'who' questions. When asked 'who' questions about her closest companions, her responses were different for each individual but consistent over repeated questions about each individual.

Human children begin by using personal pronouns and self-referents in their speech at about the same time that they begin to recognise themselves in mirrors. Similarly, at the same age that she began to exhibit self-directed behaviours in front of a mirror, Koko acquired the signs for 'me', 'mine', 'myself', 'you', and proper nouns, including 'Koko'. The gorillas have also demonstrated self-awareness in their ability to describe themselves as the same as or different from others. For example, research assistant Maureen Sheehan questioned Koko about the differences between gorillas and humans:

MS: What's the difference between you and me?
K: HEAD.
MS: And how are our heads different?
Koko beats on her head with her open hands quite hard, harder than a human would ever do.
MS: What else is different between us?

Table 3 Koko's responses to self-identity questions

Test session	First Response	Second Response
'Who are you?'		
1	ME GORILLA NIPPLES TICKLE	NIPPLES GORILLA
2	POLITE-KOKO KOKO NUT NUT POLITE	POLITE DEVIL HAIR HEAD KOKO SWEET BAD
3	KOKO POLITE ME THIRSTY	GORILLA ME
4	POLITE ME THIRSTY FEEL KOKO-LOVE	POLITE POLITE-KOKO KOKO FEEL THIRSTY SORRY
5	KOKO POLITE SORRY GOOD	KOKO; PLEASE GIMME BROW-WIPER
'Who is that?' (to mirror)		
5	ME THERE KOKO GOOD TEETH GOOD	

Koko moves her hand up to her stomach, a gesture resembling BLANKET.
MS: Do you mean something about your stomach?
K: STOMACH GOOD THAT.
MS: Oh, but what were you saying about blanket, different?
Koko moves her hands up and down her torso, then pulls the hair on her belly. Maureen interprets these new gestures as meaning 'body hair'.
MS: Now can you name something the same?
K: EYE.
MS: Yes, that's right, we both have eyes.
K: LOVE.

Michael was once asked a related question: 'What makes you different from your cat?' He responded, 'GORILLA ME MIKE'.

Self-awareness is also shown by the gorillas when they talk about themselves in situations removed in space or time from the current one. This capacity for displacement is clearly and repeatedly demonstrated in conversations such as the following:

10 July 1984, with Francine Patterson, six days after Koko's birthday
F: What happened on your birthday?
K: SLEEP EAT.
F: Didn't something special happen on your birthday?
K: OLD GORILLA.
F: Yes, you're thirteen years old and big.

Michael began at an early age to tell stories about past events. In 1979, when the project was still located on the Stanford University campus, the gorillas witnessed a violent, screaming argument between a red-haired woman and a research assistant at a nearby laboratory building. The woman apparently hit the assistant and had to be subdued at gunpoint by the police. When Michael's companion Barbara Weller arrived, knowing nothing of the incident, she found him at the window . . .

M: GIRL.
Michael stares out of the window for a few more minutes.
M: KNOW HIT-IN-MOUTH.
B: Know hit-in-mouth?
M: HIT-IN-MOUTH RED BITE.
B: Why are you signing hit-in-mouth?
M: KNOW.
B: Who do you want to hit-in-mouth?
M: HAIR GIRL RED.
B: What? Red hair girl?
M: LIP [The gorillas use LIP to mean 'woman'.]
B: Lip?
M: LIP LIP LIP BIG-TROUBLE.

Michael Lewis[11] suggests that there are certain emotional states that only the self-aware can have. One such state is embarrassment. In order to be embarrassed animals must be capable of reflecting on their own behaviour and comparing it to standards set by society or themselves. Koko seems embarrassed when her companions notice that she is signing to herself, especially when the signing involves her dolls. One incident recorded when Koko was five years old provides an example. Her companion observed her creating what appeared to be an imaginary social situation between two gorilla dolls. She signed, 'BAD, BAD' while looking at one doll, and 'KISS' while looking at the other. Next, she signed, 'CHASE, TICKLE', hit the two dolls together, and then wrestled with them and signed, 'GOOD GORILLA, GOOD GOOD'. At this point she noticed that she was being watched and abruptly put the dolls down.

There is no reason to think that Koko and Michael are significantly different from other gorillas in their inherent linguistic capacities, self-awareness or other mental abilities. They are two individuals selected more or less at random from the total population of gorillas, and the circumstances of their first few years were very different. So it is fair to assume that they are representative of their species. Nor is there reason to consider them essentially different from other gorillas because of their experience with human language. Indeed, a few zoo gorillas who have been exposed informally to sign language have shown that they, too, can learn signs, even later in life and without intensive teaching. By teaching sign language to Koko and Michael we have not imposed an artificial system on them, but rather have built upon their existing system to provide a jointly understood vocabulary for mutual exchange.

Detailed observation and analysis of the communicative gestures used by 'uninstructed' gorillas in a zoo group indicate that their own gestural communication system is much more complex then previously thought.[12] This ongoing study involves analysing a videotape compilation of these gestures and classifying them according to context and apparent functions. The gorillas have been observed to use communicative gestures in the following contexts: play invitation, anticipatory play reaction, play inhibition, indication of play location and action, sexual activities, agonistic interaction, group movement, body positioning and solitary play. One type of gesture involves touching to position the body of another, usually in the context of sexual activity. Another significant type of gesture uses the hands for deception, for example to hide a 'playface' grin in order to alter the signal being received by another gorilla. So far, over forty apparently discrete and meaningful gestural types have been identified in one five-member population. Gorillas in this particular group have been observed using conversational strings of up to eight gestures, and there seems to be an element of request and response in their dialogues.

While we are a long way from any comprehensive understanding of natural gorilla communication, it is clear that non-signing gorillas use gestures to communicate with one another. Field researchers may not have always recognised the significance of semantic gestures used by free-living gorillas, because they were unfamiliar with the gorillas' communicative habits or with gestural communication in general, or because the presence of human observers inhibits the gorillas' normal behaviour. Recognition of semantically significant gestures and sounds becomes easier as we become more familiar with gorillas as communicators.

Perhaps our most interesting findings relate to how astonishingly like us gorillas are – or how like them we are. But the striking similarities between gorillas and humans are hardly surprising in light of the most recent studies of our genetic kinship. The scientific classification of living organisms is based on the apparent similarities between those organisms. Within the order Primates, human beings have always been set apart in a separate family. More recent studies involving comparisons of chromosomes and analysis of DNA leave little doubt that apes and humans should be classed together in the family *Hominidae*. Some researchers now propose that humans, gorillas and chimpanzees also belong in the same subfamily, though the arrangement within this subfamily is still to be determined.[13]

Through what they have taught us about gorillas, Koko and Michael are helping to change the way we view the world. They force us to re-examine the ways we think about other animals. With an emotional and expressive range far greater than previously believed, they have revealed a lively and sure awareness of themselves as individuals. Asked to categorise herself, Koko declared 'FINE ANIMAL GORILLA'. Indeed. Fine animal-persons, gorillas.

Notes

1. Additional information about the work of the Gorilla Foundation with Koko and Michael can be found in: F. G. Patterson, 'The gestures of a gorilla: language acquisition in another pongid', *Brain and Language*, vol. 5 (1978) pp. 72–97; F. Patterson, 'Conversations with a gorilla', *National Geographic*, vol. 154, no. 4 (1978) pp. 438–65; F. Patterson and E. Linden, *The Education of Koko* (Holt, Rinehart and Winston, New York, 1981); F. Patterson, C. H. Patterson, and D. K. Brentari, 'Language in child, chimp, and gorilla', *American Psychologist*, vol. 42, no. 3 (1987) pp. 270–2; F. Patterson, J. Tanner, and N. Mayer, 'Pragmatic analysis of gorilla utterances: early communicative development in the gorilla Koko', *Journal of Pragmatics*, vol. 12, no.1 (1988) pp. 35–55.

2. F. G. Patterson, 'Linguistic capabilities of a young lowland gorilla', in F. C. Peng (ed.), *Sign Language and Language Acquisition in Man and Ape: New Dimensions in Comparative Pedolinguistics* (Westview Press, Boulder, CO, 1978), pp. 161–201.

3. F. G. Patterson, 'Linguistic capabilities of a lowland gorilla' (Ph.D. dissertation, Stanford University, 1979. University Microfilms International no. 79–172–69, Abstract in *Dissertation Abstracts International*, August 1979, 40–B, 2).

4. F. G. Patterson and R. H. Cohn, 'Language acquisition by a lowland gorilla: Koko's first ten years of vocabulary development', *Word*, vol. 41, no. 2 (1990) pp. 97–143.

5. H. Gardner, 'Metaphors and modalities: how children project polar adjectives onto diverse domains', *Child Development*, vol. 45 (1974) pp. 84–91.

6. F. G. Patterson, 'Innovative uses of language by a gorilla: a case study', in K. Nelson (ed.), *Children's Language*, vol. 2 (Gardner, New York, 1980), pp. 497–561.

7. D. Hancocks, 'Gorillas go natural', *Animal Kingdom*, vol. 86, no.1 (1983) pp. 10–16.

8. A. Caplan, 'Moral community and the responsibility of scientists', *Acta Physiologica Scandinavica*, vol. 128 (1986) p. 554.

9. R. W. Wolman, W. C. Lewis and M. King, 'The development of the language of emotions: conditions of emotional arousal', *Child Development*, vol. 42 (1971) pp. 1288–93.

10. F. Patterson and R. Cohn, 'Self-recognition and self-awareness in lowland gorillas', in S. T. Parker, M. L. Boccia and R. Mitchell (eds), *Self-awareness in Animals and Humans* (Cambridge University Press, Cambridge, in press).

11. M. Lewis, 'Origins of self-knowledge and individual differences in early self-recognition', in A. Greenwald and T. Suls (eds), *Psychological Perspective on the Self*, vol. 3 (1986) pp. 55–78.

12. F. Patterson and J. Tanner, 'Gestural communication in captive gorillas', paper presented at the American Society of Primatologists meeting, University of California, Davis, July 1990.

13. J. Yunis and O. Prakash, 'The origins of man: a chromosomal pictorial legacy', *Science*, vol. 215 (1982) pp. 1525–9; B. F. Koop, M. Goodman, P. Xu, K. Chan and J. L. Slightom, 'Primate (eta)-globin DNA sequences and man's place among the great apes', *Nature*, vol. 319 (1986) pp. 234–7.

III

Similarity and Difference

7

Gaps in the Mind

RICHARD DAWKINS

Richard Dawkins did his doctoral work at the University of Oxford with the Nobel Prize winning ethologist Niko Tinbergen. He is now a lecturer in zoology at Oxford University, and a fellow of New College. His first book, The Selfish Gene, *became an international bestseller and established Dawkins's reputation as a consummate interpreter of modern evolutionary theory with a knack for reaching a wide audience – a reputation that was further enhanced in 1986 by the publication of* The Blind Watch-maker. *Dawkins's gift for vivid exposition of scientific ideas is evident in the essay that follows, in which he challenges our tendency to put a gap between ourselves and the other great apes.*

Sir,
You appeal for money to save the gorillas. Very laudable, no doubt. But it doesn't seem to have occurred to you that there are thousands of *human* children suffering on the very same continent of Africa. There'll be time enough to worry about gorillas when we've taken care of every last one of the kiddies. Let's get our priorities right, *please!*

This hypothetical letter could have been written by almost any well-meaning person today. In lampooning it, I don't mean to imply that a good case could not be made for giving human children priority. I expect it could, and also that a good case could be made the other way. I'm only trying to point the finger at the *automatic*, unthinking nature of the speciesist double standard. To many people it is simply self-evident, *without any discussion*, that humans are entitled to special treatment. To see this, consider the following variant on the same letter:

Sir,
You appeal for money to save the gorillas. Very laudable, no doubt. But it doesn't seem to have occurred to you that there are

thousands of *aardvarks* suffering on the very same continent of Africa. There'll be time enough to worry about gorillas when we've saved every last one of the aardvarks. Let's get our priorities right, *please!*

This second letter could not fail to provoke the question: What's so special about aardvarks? A good question, and one to which we should require a satisfactory answer before we took the letter seriously. Yet the first letter, I suggest, would not for most people provoke the equivalent question: What's so special about humans? As I said, I don't deny that this question, unlike the aardvark question, very probably has a powerful answer. All that I am criticising is an unthinking failure to realise in the case of humans that the question even arises.

The speciesist assumption that lurks here is very simple. Humans are humans and gorillas are animals. There is an unquestioned yawning gulf between them such that the life of a single human child is worth more than the lives of all the gorillas in the world. The 'worth' of an animal's life is just its replacement cost to its owner – or, in the case of a rare species, to humanity. But tie the label *Homo sapiens* even to a tiny piece of insensible, embryonic tissue, and its life suddenly leaps to infinite, uncomputable value.

This way of thinking characterises what I want to call the discontinuous mind. We would all agree that a six-foot woman is tall, and a five-foot woman is not. Words like 'tall' and 'short' tempt us to force the world into qualitative classes, but this doesn't mean that the world really is discontinuously distributed. Were you to tell me that a woman is five feet nine inches tall, and ask me to decide whether she should therefore be called tall or not, I'd shrug and say 'She's five foot nine, doesn't that tell you what you need to know?' But the discontinuous mind, to caricature it a little, would go to court (probably at great expense) to decide whether the woman was tall or short. Indeed, I hardly need to say caricature. For years, South African courts have done a brisk trade adjudicating whether particular individuals of mixed parentage count as white, black or coloured.

The discontinuous mind is ubiquitous. It is especially influential when it afflicts lawyers and the religious (not only are all judges lawyers; a high proportion of politicians are too, and all politicians have to woo the religious vote). Recently, after giving a public lecture, I was cross-examined by a lawyer in the audience. He brought the full weight of his legal acumen to bear on a nice point of evolution. If species A evolves into a later species B, he reasoned closely, there must come a point when a mother belongs to the old species A and her child belongs to the new species B. Members of different species cannot interbreed with one another. I put it to you, he went on, that a child could hardly be so different from its parents that it could not interbreed with their kind. So, he wound up triumphantly, isn't this a fatal flaw in the theory of evolution?

But it is we that choose to divide animals up into discontinuous species. On the evolutionary view of life there must have been intermediates, even though, conveniently for our naming rituals, they are usually extinct: usually, but not always. The lawyer would be surprised and, I hope, intrigued by so-called 'ring species'. The best-known case is herring gull versus lesser black-backed gull. In Britain these are clearly distinct species, quite different in colour. Anybody can tell them apart. But if you follow the population of herring gulls westward round the North Pole to North America, then via Alaska across Siberia and back to Europe again, you will notice a curious fact. The 'herring gulls' gradually become less and less like herring gulls and more and more like lesser black-backed gulls until it turns out that our European lesser black-backed gulls actually are the other end of a ring that started out as herring gulls. At every stage around the ring, the birds are sufficiently similar to their neighbours to interbreed with them. Until, that is, the ends of the continuum are reached, in Europe. At this point the herring gull and the lesser black-backed gull never interbreed, although they are linked by a continuous series of interbreeding colleagues all the way round the world. The only thing that is special about ring species like these gulls is that the intermediates are still alive. *All* pairs of related species are potentially ring species. The intermediates must have lived once. It is just that in most cases they are now dead.

The lawyer, with his trained discontinuous mind, insists on placing individuals firmly in this species or that. He does not allow for the possibility that an individual might lie half-way between two species, or a tenth of the way from species A to species B. Self-styled 'pro-lifers', and others that indulge in footling debates about exactly when in its development a foetus 'becomes human', exhibit the same discontinuous mentality. It is no use telling these people that, depending upon the human characteristics that interest you, a foetus can be 'half human' or 'a hundredth human'. 'Human', to the discontinuous mind, is an absolute concept. There can be no half measures. And from this flows much evil.

The word 'apes' usually means chimpanzees, gorillas, orang-utans, gibbons and siamangs. We admit that we are like apes, but we seldom realise that we *are* apes. Our common ancestor with the chimpanzees and gorillas is much more recent than their common ancestor with the Asian apes – the gibbons and orang-utans. There is no natural category that includes chimpanzees, gorillas and orang-utans but excludes humans. The artificiality of the category 'apes', as conventionally taken to exclude humans, is demonstrated by Figure 7.1. This family tree shows humans to be in the thick of the ape cluster; the artificiality of the conventional category 'ape' is shown by the stippling.

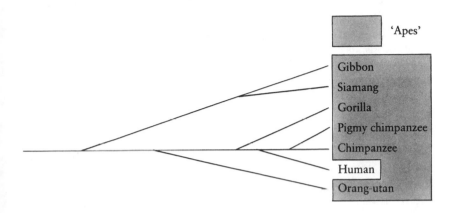

Figure 7.1

In truth, not only are we apes, we are African apes. The category 'African apes', if you don't arbitrarily exclude humans, is a natural one. The stippled area in Figure 7.2 doesn't have any artificial 'bites' taken out of it.

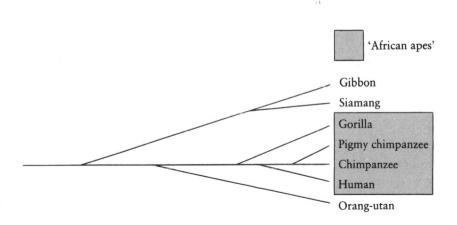

Figure 7.2

'Great apes', too, is a natural category only so long as it includes humans. We are great apes. All the great apes that have ever lived, including ourselves, are linked to one another by an unbroken chain of parent-child bonds. The same is true of all animals and plants that have ever lived, but there the distances involved are much greater. Molecular evidence suggests that our common ancestor with chimpanzees lived, in Africa, between five and seven million years ago, say half a million generations ago. This is not long by evolutionary standards.

Happenings are sometimes organised at which thousands of people hold hands and form a human chain, say from coast to coast of the United States, in aid of some cause or charity. Let us imagine setting one up along the equator, across the width of our home continent of Africa. It is a special kind of chain, involving parents and children, and we will have to play tricks with time in order to imagine it. You stand on the shore of the Indian Ocean in southern Somalia, facing north, and in your left hand you hold the right hand of your mother. In turn she holds the hand of her mother, your grandmother. Your grandmother holds her mother's hand, and so on. The chain wends its way up the beach, into the arid scrubland and westwards on towards the Kenya border.

How far do we have to go until we reach our common ancestor with the chimpanzees? It is a surprisingly short way. Allowing one yard per person, we arrive at the ancestor we share with chimpanzees in under 300 miles. We have hardly started to cross the continent; we are still not half way to the Great Rift Valley. The ancestor is standing well to the east of Mount Kenya, and holding in her hand an entire chain of her lineal descendants, culminating in you standing on the Somali beach.

The daughter that she is holding in her right hand is the one from whom we are descended. Now the arch-ancestress turns eastward to face the coast, and with her left hand grasps her other daughter, the one from whom the chimpanzees are descended (or son, of course, but let's stick to females for convenience). The two sisters are facing one another, and each holding their mother by the hand. Now the second daughter, the chimpanzee ancestress, holds her daughter's hand, and a new chain is formed, proceeding back towards the coast. First cousin faces first cousin, second cousin faces second cousin, and so on. By the time the folded-back chain has reached the coast again, it consists of modern chimpanzees. You are face to face with your chimpanzee cousin, and you are joined to her by an unbroken chain of mothers holding hands with daughters. If you walked up the line like an inspecting general – past *Homo erectus*, *Homo habilis*, perhaps *Australopithecus afarensis* – and down again the other side (the intermediates on the chimpanzee side are unnamed because, as it happens, no fossils have been found), you would nowhere find any sharp discontinuity. Daughters would resemble mothers just as much (or as little) as they always do. Mothers would love daughters, and feel affinity with them, just as they always do. And this hand-in-hand continuum, joining us seamlessly to chimpanzees, is so short that it barely makes it past the hinterland of Africa, the mother continent.

Our chain of African apes, doubling back on itself, is in miniature like the ring of gulls round the pole, except that the intermediates happen to be dead. The point I want to make is that, as far as morality is concerned, it should be incidental that the intermediates are dead. What if they were not? What if a clutch of intermediate types had survived, enough to link us to modern chimpanzees by a chain, not just of handholders, but of interbreeders? Remember the song, 'I've danced with a man, who's danced with a girl, who's danced with the Prince of Wales'? We can't (quite) interbreed with modern chimpanzees, but we'd need only a handful of intermediate types to be able to sing: 'I've bred with a man, who's bred with a girl, who's bred with a chimpanzee.'

It is sheer luck that this handful of intermediates no longer exists. ('Luck' from some points of view: for myself, I should love to meet them.) But for this chance, our laws and our morals would be very different. We need only discover a single survivor, say a relict *Australopithecus* in the Budongo Forest, and our precious system of norms and ethics would come crashing about our ears. The boundaries with which we segregate our world would be all shot to pieces. Racism would blur with speciesism in obdurate and vicious confusion. Apartheid, for those that believe in it, would assume a new and perhaps a more urgent import.

But why, a moral philosopher might ask, should this matter to us? Isn't it only the discontinuous mind that wants to erect barriers anyway? So what if, in the continuum of all apes that have lived in Africa, the survivors happen to leave a convenient gap between *Homo* and *Pan*? Surely we should, in any case, not base our treatment of animals on whether or not we can interbreed with them. If we want to justify double standards – if society agrees that people should be treated better than, say, cows (cows may be cooked and eaten, people may not) – there must be better reasons than cousinship. Humans may be taxonomically distant from cows, but isn't it more important that we are brainier? Or better, following Jeremy Bentham, that humans can suffer more – that cows, even if they hate pain as much as humans do (and why on earth should we suppose otherwise?), do not know what is coming to them? Suppose that the octopus lineage had happened to evolve brains and feelings to rival ours; they easily might have done. The mere possibility shows the incidental nature of cousinship. So, the moral philosopher asks, why emphasise the human/chimp continuity?

Yes, in an ideal world we probably should come up with a better reason than cousinship for, say, preferring carnivory to cannibalism.

But the melancholy fact is that, at present, society's moral attitudes rest almost entirely on the discontinuous, speciesist imperative.

Figure 7.3 Hypothetical computer-generated image of what an intermediate between a human and a chimpanzee face might look like. (After Nancy Burston and David Kramlich, from C. A. Pickover, *Computers and the Imagination: Visual Adventures Beyond the Edge* (Alan Sutton, Stroud, 1991).)

This arresting picture is hypothetical. But I can assert, without fear of contradiction, that if somebody succeeded in breeding a chimpanzee/human hybrid the news would be earth-shattering. Bishops would bleat, lawyers would gloat in anticipation, conservative politicians would thunder, socialists wouldn't know where to put their barricades. The scientist that achieved the feat would be drummed out of politically correct common-rooms; denounced in pulpit and gutter press; condemned, perhaps, by an Ayatollah's *fatwah*. Politics would never be the same again, nor would theology, sociology, psychology or most branches of philosophy. The world that would be so shaken, by such an incidental event as a hybridisation, is a speciesist world indeed, dominated by the discontinuous mind.

I have argued that the discontinuous gap between humans and 'apes' that we erect in our minds is regrettable. I have also argued that, in any case, the present position of the hallowed gap is arbitrary, the result of evolutionary accident. If the contingencies of survival and extinction had been different, the gap would be in a different place. Ethical principles that are based upon accidental caprice should not be respected as if cast in stone.

Nevertheless, it must be conceded that this book's proposal to admit great apes to the charmed circle of human privilege stands square in the discontinuous tradition. Albeit the gap has moved, the fundamental question is still 'Which side of the gap?' Regrettable as this is, as long as our social mores are governed by discontinuously minded lawyers and theologians, it is premature to advocate a quantitative, continuously distributed morality. Accordingly, I support the proposal for which this book stands.

8

The Third Chimpanzee

JARED DIAMOND

Jared Diamond trained in physiology and is now professor of physiology at the University of California, Los Angeles; but he has also made fundamental contributions to ecology. On one of his many trips to the mountains of New Guinea, he rediscovered a long-lost species of bowerbird. 'The Third Chimpanzee' is a slightly abridged version of the opening chapter of his recent much-acclaimed book The Rise and Fall of the Third Chimpanzee *(Harper Collins, New York, and Radius, London, 1991) and is reprinted by kind permission of the author and publishers. © Jared Diamond, 1991.*

The next time that you visit a zoo, make a point of walking past the ape cages. Imagine that the apes had lost most of their hair, and imagine a cage nearby holding some unfortunate people who had no clothes and couldn't speak but were otherwise normal. Now try guessing how similar those apes are to ourselves genetically. For instance, would you guess that a chimpanzee shares 10, 50 or 99 per cent of its genes with humans?

Then ask yourself why those apes are on exhibition in cages, and why other apes are being used for medical experiments, while it is not permissible to do either of those things to humans. Suppose it turned out that chimps shared 99.9 per cent of their genes with us, and that the important differences between humans and chimps were due to just a few genes. Would you still think it is okay to put chimps in cages and to experiment on them? Consider those unfortunate mentally impaired people who have much less capacity to solve problems, to care for themselves, to communicate, to engage in social relationships and to feel pain, than do apes. What is the logic that forbids medical experiments on those people, but not on apes?

You might answer that apes are 'animals', while humans are humans, and that is enough. An ethical code for treating humans should not be extended to an 'animal', no matter what percentage of its genes it shares with us, and no matter what its capacity for social relationships or for feeling pain. That is an arbitrary but at least self-consistent answer that cannot be lightly dismissed. In that case, learning more about our ancestral relationships will not have any ethical consequences, but it will still satisfy our intellectual curiosity to understand where we come from. Every human society has felt a deep need to make sense of its origins, and has answered that need with its own story of the Creation. What follows is the creation story of our time.

For centuries it has been clear approximately where we fit into the animal kingdom. We are obviously mammals, the group of animals characterised by having hair, nursing their young and other features. Among mammals we are obviously primates, the group of mammals including monkeys and apes. We share with other primates numerous traits lacking in most other mammals, such as flat fingernails and toenails rather than claws, hands for gripping, a thumb that can be opposed to the other four fingers and a penis that hangs free rather than being attached to the abdomen. By the second century AD, the Greek physician Galen had deduced our approximate place in nature correctly when he dissected various animals and found that a monkey was 'most similar to man in viscera, muscles, arteries, veins, nerves and in the form of bones'.

It is also easy to place us within the primates, among which we are obviously more similar to apes than to monkeys. To name only one of the most visible signs, monkeys sport tails, which, like the apes, we lack. It is also clear that gibbons, with their small size and very long arms, are the most distinctive apes, and that orang-utans, chimpanzees, gorillas and humans are all more closely related to each other than any of them is to gibbons. But to go further with our relationships proves unexpectedly difficult. It has provoked an intense scientific debate, which revolves around three questions including the one that I posed in the first paragraph in this chapter:

1. What is the detailed family tree of relationships among humans, the living apes and extinct ancestral apes? For example, which of the living apes is our closest relative?
2. When did we and that closest living relative, whichever ape it is, diverge from a common ancestor?
3. What fraction of our genes do we share with that closest living relative?

At first, it would seem natural to assume that comparative anatomy had already solved the first of those three questions. We look especially like chimpanzees and gorillas, but differ from them in obvious features such as our larger brains, upright posture and much sparser body hair, as well as in many more subtle points. However, on closer examination these anatomical facts are not decisive. Depending on what anatomical characters one considers most important and how one interprets them, biologists differ on whether we are most closely related to the orang-utan (the minority view), with chimps and gorillas having branched off our family tree before we split off from orang-utans, or whether we are instead closest to chimps and gorillas (the majority view), with the ancestors of orang-utans having gone their separate way earlier.

Within the majority, most biologists have thought that gorillas and chimps are more like each other than either is like us, implying that we branched off before the gorillas and chimps diverged from each other. This conclusion reflects the common-sense view that chimps and gorillas can be lumped in a category termed 'apes', while we are something different. However, it is also conceivable that we look distinct only because chimps and gorillas have not changed much since we diverged from our common ancestor, while we have changed greatly in a few important and highly visible features, such as upright posture and brain size. In that case, humans might be most similar to gorillas, or humans might be most similar to chimps, or humans and gorillas and chimps might be roughly equidistant from each other, in overall genetic make-up.

Hence, anatomists have continued to argue about the first question, the details of our family tree. Whichever tree one prefers, anatomical studies by themselves tell us nothing about the second and third questions, our time of divergence and genetic distance from apes. Perhaps fossil evidence might in principle solve the questions of the correct ancestral tree and of dating, though not the question of genetic distance. If we had abundant fossils, we might hope to find a series of dated proto-human fossils and another series of dated proto-chimp fossils converging on a common ancestor around ten million years ago, converging in turn on a series of proto-gorilla fossils twelve million years ago. Unfortunately, that hope for insight from the fossil record has also been frustrated, because almost no ape fossils of any sort have been found for the crucially relevant period between five and fourteen million years ago in Africa.

The solution to these questions about our origins came from an unexpected direction: molecular biology as applied to bird taxonomy. About thirty years ago, molecular biologists began to realise that the chemicals of which plants and animals are composed might provide 'clocks' by which to measure genetic distances and to date times of evolutionary divergence. The idea is as follows. Suppose there is some class of molecules that occurs in all species, and whose particular structure in each species is genetically determined. Suppose further that that structure changes slowly over the course of millions of years because of genetic mutations, and that the rate of change is the same in all species. Two species derived from a common ancestor would start off with identical forms of the molecule, which they inherited from that ancestor, but mutations would then occur independently and produce structural changes between the molecules of the two species. The two species' versions of the molecule would gradually diverge in structure. If we knew how many structural changes now occur on the average every million years, we could then use the difference today in the molecule's structure between any two related animal species as a clock, to calculate how much time had passed since the species shared a common ancestor.

For instance, suppose one knew from fossil evidence that lions and tigers diverged five million years ago. Suppose the molecule in lions were 99 per cent identical in structure to the corresponding molecule in tigers and differed only by 1 per cent. If one then took a pair of species of unknown fossil history and found that the molecule differed by 3 per cent between those two species, the molecular clock would say that they had diverged three times five million, or fifteen million, years ago.

Neat as this scheme sounds on paper, testing whether it succeeds in practice has cost biologists much effort. Four things had to be done before molecular clocks could be applied: find the best molecule; find a quick way of measuring changes in its structure; prove that the clock runs steady (that is, that the molecule's structure really does evolve at the same rate among all the species that one is studying); and measure what that rate is.

Molecular biologists had worked out the first two of these problems by around 1970. The best molecule proved to be deoxyribonucleic acid (abbreviated to DNA), the famous substance whose structure James Watson and Francis Crick showed to consist of a double helix, thereby revolutionising the study of genetics. DNA is made up of two complementary and extremely long chains, each made up of four types of small molecules whose sequence within the chain carries all the genetic information transmitted from parents to offspring. A quick method of measuring changes in DNA structure is to mix the DNA from two species, then to measure by how many degrees of temperature the melting-point of the mixed (hybrid) DNA is reduced below the melting-point of pure DNA from a single species. Hence the method is generally referred to as DNA hybridisation. As it turns out, a melting-point lowered by one degree Celsius (abbreviated: delta $T = 1°C$) means that the DNAs of the two species differ by roughly 1 per cent.

In the 1970s, most molecular biologists and most taxonomists had little interest in each other's work. Among the few taxonomists who appreciated the potential power of the new DNA hybridisation technique was Charles Sibley, an ornithologist then serving as Professor of Ornithology and Director at Yale's Peabody Museum of Natural History. Bird taxonomy is a difficult field because of the severe anatomical constraints imposed by flight. There are only so many ways to design a bird capable, say, of catching insects in mid-air, with the result that birds of similar habits tend to have very similar anatomies, whatever their ancestry. For example, American vultures look and behave much like Old World vultures, but biologists have come to realise that the former are related to storks, the latter to hawks, and that their resemblances result from their common lifestyle. Frustrated by the shortcomings of traditional methods for deciphering bird relationships, Sibley and Jon Ahlquist turned in 1973 to the DNA clock, in the most massive application to date of the methods of molecular biology to taxonomy. Not until 1980 were Sibley and Ahlquist ready to begin publishing their results, which eventually came to encompass applying the DNA clock to about 1,700 bird species – nearly one-fifth of all living birds.

While Sibley's and Ahlquist's achievement was a monumental one, it initially caused much controversy because so few other scientists possessed the blend of expertise required to understand it. Here are typical reactions I heard from my scientist friends:

'I'm sick of hearing about that stuff. I no longer pay attention to anything those guys write' (an anatomist).

'Their methods are OK, but why would anyone want to do something so boring as all that bird taxonomy?' (a molecular biologist).

'Interesting, but their conclusions need a lot of testing by other methods before we can believe them' (an evolutionary biologist).

'Their results are The Revealed Truth, and you'd better believe it' (a geneticist).

My own assessment is that the last view will prove to be the most nearly correct one. The principles on which the DNA clock rests are unassailable; the methods used by Sibley and Ahlquist are state-of-the-art; and the internal consistency of their genetic-distance measurements from over 18,000 hybrid pairs of bird DNA testifies to the validity of their results.

Just as Darwin had the good sense to marshal his evidence for variation in barnacles before discussing the explosive subject of human variation, Sibley and Ahlquist similarly stuck to birds for most of the first decade of their work with the DNA clock. Not until 1984 did they publish their first conclusions from applying the same DNA methods to human origins, and they refined their conclusions in later papers. Their study was based on DNA from humans and from all of our closest relatives: the common chimpanzee, pygmy chimpanzee, gorilla, orang-utan, two species of gibbons and seven species of Old World monkeys. Figure 8.1 summarises the results.

As any anatomist would have predicted, the biggest genetic difference, expressed in a big DNA melting-point lowering, is between monkey DNA and the DNA of humans or of any ape. This simply puts a number on what everybody has agreed ever since apes first became known to science: that humans and apes are more closely related to each other than either are to monkeys. The actual statistic is that monkeys share 93 per cent of their DNA structure with humans and apes, and differ in 7 per cent.

Equally unsurprising is the next biggest difference, one of 5 per cent between gibbon DNA and the DNA of other apes or humans. This too confirms the accepted view that gibbons are the most distinct apes, and that our affinities are instead with gorillas, chimpanzees and orang-utans. Among those latter three groups of apes, most recent anatomists have considered the orang-utan as somewhat separate, and that conclusion too fits the DNA evidence: a difference of 3.6 per cent between orang-utan DNA and that of humans, gorillas and chimpanzees. Geography confirms that the latter three species parted from gibbons and orang-utans quite some time ago: living and fossil gibbons and orang-utans are confined to Southeast Asia, while living gorillas and chimpanzees plus early fossil humans are confined to Africa.

At the opposite extreme but equally unsurprising, the most similar DNAs are those of common chimpanzees and pygmy chimpanzees, which are 99.3 per cent identical and differ by only 0.7 per cent. So similar are these two chimp species in appearance that it was not until 1929 that anatomists even bothered to give them separate names. Chimps living on the equator in central Zaïre rate the name 'pygmy chimps' because they are on average slightly smaller (and have more slender builds and longer legs) than the widespread 'common chimps' ranging across Africa just north of the equator. However, with the increased knowledge of chimp behaviour acquired in recent years, it has become clear that the modest anatomical differences between pygmy

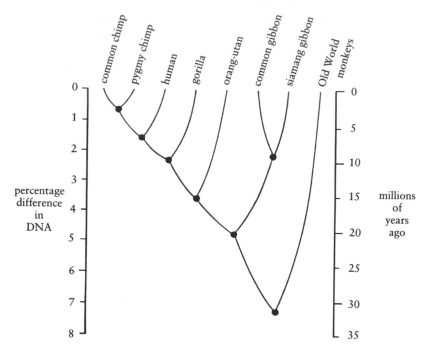

Figure 8.1   Trace back each pair of modern higher primates to the black dot connecting them. The numbers to the left then give the percentage difference between the DNAs of those modern primates, while the numbers to the right give the estimated number of millions of years since they diverged from a common ancestor. For example, the common and pygmy chimps differ in about 0.7% of their DNA and diverged about three million years ago; we differ in about 1.6% of our DNA from both species of chimps and diverged from their common ancestor about seven million years ago; gorillas differ in about 2.3% of their DNA from us and from chimps and diverged from the common ancestor leading to us and the two chimps about ten million years ago.

and common chimps mask considerable differences in reproductive biology. Unlike common chimps, but like ourselves, pygmy chimps assume a wide variety of positions for copulation, including face-to-face; copulation can be initiated by either sex, not just by the male; females are sexually receptive for much of the month, not just for a briefer period in mid-month; and there are strong bonds among females or between males and females, not just among males. Evidently, those few genes (0.7 per cent) that differ between pygmy and common chimps have big consequences for sexual physiology and roles. That same theme – a small percentage of gene differences having great consequences – will recur later in this chapter with regard to the gene differences between humans and chimps.

In all the cases that I have discussed so far, anatomical evidence of relationships was already convincing, and the DNA-based conclusions confirmed what the anatomists had already concluded. But DNA was also able to resolve the problem at which anatomy had failed – the relationships between humans, gorillas and chimpanzees. As Figure 8.1 shows, humans differ from both common chimps and pygmy chimps in about 1.6 per cent of their (our) DNA, and share 98.4 per cent. Gorillas differ somewhat more, by about 2.3 per cent, from us and from both of the chimps.

Let us pause to let some of the implications of these momentous numbers sink in.

The gorilla must have branched off from our family tree slightly before we separated from the common and pygmy chimpanzees. The chimpanzees, not the gorilla, are our closest relatives. Put another way, the chimpanzees' closest relative is not the gorilla but the human. Traditional taxonomy has reinforced our anthropocentric tendenciesby claiming to see a fundamental dichotomy between mighty man, standing alone on high, and the lowly apes all together in the abyss of bestiality. Now future taxonomists may see things from the chimpanzees' perspective: a weak dichotomy between slightly higher apes (the *three* chimpanzees, including the 'human chimpanzee') and slightly lower apes (gorilla, orang-utan, gibbon). The traditional distinction between 'apes' (defined as chimps, gorillas, etc.) and humans misrepresents the facts.

The genetic distance (1.6 per cent) separating us from pygmy or common chimps is barely double that separating pygmy from common chimps (0.7 per cent). It is less than that between two species of gibbons (2.2 per cent), or between such closely related North American bird species as red-eyed vireos and white-eyed vireos (2.9 per cent), or between such closely related and hard-to-distinguish European bird species as willow warblers and chiffchaffs (2.6 per cent). The remaining 98.4 per cent of our genes are just normal chimp genes. For example, our principal haemoglobin, the oxygen-carrying protein that gives blood its red colour, is identical in all 287 units with chimp haemoglobin. In this respect as in most others, we are just a third species of chimpanzee, and what is good enough for common and pygmy chimps is good enough for us. Our important visible distinctions from the other chimps – our upright posture, large brains, ability to speak, sparse body hair, and peculiar sexual lives – must be concentrated in a mere 1.6 per cent of our genes.

If genetic distances between species accumulated at a uniform rate with time, they would function as a smoothly ticking clock. All that would be required to convert genetic distance into absolute time since the last common ancestor would be a calibration, furnished by a pair of species for which we know *both* the genetic distance *and* the time of divergence as dated independently by fossils. In fact, two independent calibrations are available for higher primates. On the one hand, monkeys diverged from apes between twenty-five and thirty million years ago, according to fossil evidence, and now differ in about 7.3 per cent of their DNA. On the other hand, orang-utans diverged from chimps and gorillas between twelve and sixteen million years ago and now differ in about 3.6 per cent of their DNA. Comparing these two examples, a doubling of evolutionary time, as one goes from twelve or sixteen to twenty-five or thirty million years, leads to a doubling of genetic distance (from 3.6 to 7.3 per cent of DNA). Thus, the DNA clock has ticked relatively steadily among higher primates.

With those calibrations, Sibley and Ahlquist estimated the following time scale for our evolution. Since our own genetic distance from chimps (1.6 per cent) is about half the distance of orang-utans from chimps (3.6 per cent), we must have been going our separate way for about half of the twelve to sixteen million years that orang-utans had to accumulate their genetic distinction from chimps. That is, the human and 'other chimp' evolutionary lines diverged around six to eight million years ago. By the same reasoning, gorillas parted from the common ancestor of us three chimpanzees around nine million years ago, and the pygmy and common chimps diverged around three million years ago. In contrast, when I took physical anthropology as a college freshman in 1954, the assigned textbooks said that humans diverged from apes fifteen to thirty million years ago. Thus, the DNA clock strongly supports a controversial conclusion also drawn from several other molecular clocks based on amino acid sequences of proteins, mitochondrial DNA and globin pseudogene DNA. Each clock indicates that humans have had only a short history as a species distinct from other apes, much shorter than paleontologists used to assume.

What do these results imply about our position in the animal kingdom? Biologists classify living things in hierarchical categories, each less distinct than the next: subspecies, species, genus, family, superfamily, order, class and phylum. The *Encyclopaedia Britannica* and all the biology texts on my shelf say that humans and apes belong to the same order, called Primates, and the same superfamily, called Hominoidea, but to separate families, called Hominidae and Pongidae. Whether Sibley's and Ahlquist's work changes this classification depends on one's philosophy of taxonomy. Traditional taxonomists group species into higher categories by making somewhat subjective evaluations of how important the differences between species are. Such taxonomists place humans in a separate family because of distinctive functional traits like large brain and bipedal posture and this classification would remain unaffected by measures of genetic distance.

However, another school of taxonomy, called cladistics, argues that classification should be objective and uniform, based on genetic distance or times of divergence. All taxonomists agree now that red-eyed and white-eyed vireos belong together in the genus *Vireo*, willow warblers and chiffchaffs in the genus *Phylloscopus*, the various species of gibbons in the genus *Hylobates*. Yet the members of each of these pairs of species are genetically more distant from each other than are humans from the other two chimpanzees, and diverged longer ago. On this basis, then, humans do not constitute a distinct family, nor even a distinct genus, but belong in the same genus as common and pygmy chimps. Since our genus name *Homo* was proposed first, it takes priority, by the rules of zoological nomenclature, over the genus name *Pan* coined for the 'other' chimps. Thus, there are not one but three species of genus *Homo* on Earth today: the common chimpanzee, *Homo troglodytes*; the pygmy chimpanzee, *Homo paniscus*; and the third chimpanzee or human chimpanzee, *Homo sapiens*. Since the gorilla is only slightly more distinct, it has almost equal right to be considered a fourth species of *Homo*.

Even taxonomists espousing cladistics are anthropocentric, and the lumping of humans and chimps into the same genus will undoubtedly be a bitter pill for them to swallow. There is no doubt, however, that whenever chimpanzees learn cladistics, or whenever taxonomists from outer space visit Earth to inventory its inhabitants, they will unhesitatingly adopt the new classification.

Which particular genes are the ones that differ between humans and chimps? Before we can consider this question, we need first to understand what it is that DNA, our genetic material, does.

Much or most of our DNA has no function and may just constitute 'molecular junk': that is, DNA molecules that have become duplicated or have lost former functions, and that natural selection has not eliminated from us because they do us no harm. Of our DNA that does have known functions, the main ones have to do with the long chains of amino acids called proteins. Certain proteins make up much of our body's structure (such as the proteins keratin, of hair, or collagen, of connective tissue), while other proteins, termed enzymes, synthesise and break down most of our body's remaining molecules. The sequences of the component small molecules (nucleotide bases) in DNA specify the sequence of amino acids in our proteins. Other parts of our functional DNA regulate protein synthesis.

Those of our observable features that are easiest to understand genetically are ones arising from single proteins and single genes. For instance, our blood's oxygen-carrying protein haemoglobin, already mentioned, consists of two amino acid chains, each specified by a single chunk of DNA (a single 'gene'). These two genes have no observable effects except through specifying the structure of haemoglobin, which is continued to our red blood cells. Conversely, haemoglobin's structure is totally specified by those genes. What you eat or how much you exercise may affect how much haemoglobin you make, but not the details of its structure.

That is the simplest situation, but there are also genes influencing many observable traits. For example, the fatal genetic disorder known as Tay–Sachs disease involves many behavioural as well as anatomical anomalies: excessive drooling, rigid posture, yellowish skin, abnormal head growth and other symptoms. We know in this case that all these observable effects result somehow from changes in a single enzyme specified by the Tay–Sachs gene, but we do not know exactly how. Since that enzyme occurs in many tissues of our bodies and breaks down a widespread cellular constituent, changes in that one enzyme have wide-ranging and ultimately fatal consequences. Conversely, some traits, such as your height as an adult, are influenced simultaneously by many genes and also by environmental factors (for example, your nutrition as a child).

While scientists understand well the function of numerous genes that specify known individual proteins, we know much less about the function of genes involved in more complex determinations of traits, such as most behavioural features. It would be absurd to think that human hallmarks such as art, language or aggression depend on a single gene. Behavioural differences among individual humans are obviously subject to enormous environmental influences, and what role genes play in such individual differences is a controversial question. However, for those consistent behavioural differences between chimps and humans, genetic differences are likely to be involved in those species' differences, even though we cannot yet specify the genes responsible. For instance, the ability of humans but not chimps to speak surely depends on differences in genes specifying the anatomy of the voice box and the wiring of the brain. A young chimpanzee brought up in a psychologist's home along with the psychologist's human baby of the same age still continued to look like a chimp and did not learn to talk or walk erect. But whether an individual human grows up to be fluent in English or Korean is independent of genes and dependent solely on its childhood linguistic environment, as proved by the linguistic attainments of Korean infants adopted by English-speaking parents.

With this as background, what can we say about the 1.6 per cent of our DNA that differs from chimp DNA? We know that the genes for our principal haemoglobin do not differ, and that certain other genes do exhibit minor differences. In the nine protein chains studied to date in both humans and common chimps, only five out of a total of 1,271 amino acids differ: one amino acid in a muscle protein called myoglobin, one in a minor haemoglobin chain called the delta chain and three in an enzyme called carbonic anhydrase. But we do not yet know which chunks of our DNA are responsible for the functionally significant differences between humans and chimps, such as differences in brain size, anatomy of the pelvis, voice box and genitalia, amount of body hair, female menstrual cycle, menopause and other traits. Those important differences certainly do not arise from the five amino acid differences detected to date. At present, all we can say with confidence is this: much of our DNA is junk; at least some of the 1.6 per cent that differs between us and chimps is already known to be junk; and the functionally significant differences must be confined to some as yet unidentified small fraction of 1.6 per cent.

While we do not know which particular genes are the crucial ones, there are numerous precedents for one or a few genes having a big impact. I just mentioned the many large and visible differences between Tay–Sachs patients and other people, all somehow arising from a single change in one enzyme. That is an example of differences among individuals of the same species. As for differences between related species, a good example is provided by the cichlid fishes of Africa's Lake Victoria. Cichlids are popular aquarium species, of which about 200 are confined to that one lake, where they evolved from a single ancestor within perhaps the last 200,000 years. Those 200 species differ among themselves in their food habits as much as do tigers and cows. Some graze on algae, others catch other fish, and still others variously crush snails, feed on plankton, catch insects, nibble the scales off other fish or specialise in grabbing fish embryos from brooding mother fish. Yet all those Lake Victoria cichlids differ from each other on the average by only about 0.4 per cent of their DNA studied. Thus it took even fewer genetic mutations to change a snail-crusher into a specialised baby-killer than it took to produce us from an ape.

Do the new results about our genetic distance from chimps have any broader implications, besides technical questions of taxonomic names? Probably the most important implications concern how we think about the place of humans and apes in the universe. Names are not just technical details but express and create attitudes. The new results do not specify how we *should* think about humans and apes, but, just as did Darwin's *On the Origin of Species*, they will probably influence how we *do* think, and it will probably take us many years to readjust our attitudes.

At present we make a fundamental distinction between animals (including apes) and humans, and this distinction guides our ethical code and actions. For instance, as I noted at the start of this chapter, it is considered acceptable to exhibit caged apes in zoos, but it is not acceptable to do the same with humans. I wonder how the public will feel when the identifying label on the chimp cage in the zoo reads '*Homo troglodytes*'.

I also noted earlier that it is considered acceptable to subject apes, but not humans, without their consent to lethal experiments for purposes of medical research. The motive for doing so is precisely because apes are so similar to us genetically. They can be infected with many of the same diseases as we can, and their bodies respond similarly to the disease organisms. Thus, experiments on apes offer a far better way to devise improved medical treatment for humans than would experiments on any other animals.

There is no socially accepted human analogue of medical research on animals, even though lethal experiments on humans would provide medical scientists with far more valuable information than do lethal experiments on chimps. Yet the human experiments performed by Nazi concentration camp physicians are widely viewed as one of the most abominable of all the Nazis' abominations. Why is it all right to perform such experiments on chimps?

Somewhere along the scale from bacteria to humans, we have to decide where killing becomes murder, and where eating becomes cannibalism. Most people draw those lines between humans and all other species. However, many people are vegetarians, unwilling to eat any animal (yet willing to eat plants). An increasingly vocal minority, belonging to the animal rights movement, objects to medical experiments on animals — or at least on certain animals. That movement is especially indignant at research on cats and dogs and primates, but generally silent about insects and bacteria.

If our ethical code makes a purely arbitrary distinction between humans and all other species, then we have a code based on naked selfishness devoid of any higher principle. If our code instead makes distinctions based on our superior intelligence, social relationships and capacity for feeling pain, then it becomes difficult to defend an all-or-nothing code that draws a line between all humans and all animals. Instead, different ethical constraints should apply to research on different species. Perhaps it is just our naked selfishness, re-emerging in a new disguise, that would advocate granting special rights to those animal species genetically closest to us. But an objective case, based on the considerations I have just mentioned (intelligence, social relationships, etc.) can be made that chimps and gorillas qualify for preferred ethical consideration over insects and bacteria. If there is any animal species currently used in medical research for which a total ban on medical experimentation can be justified, that species is surely the chimpanzee.

9

Common Sense, Cognitive Ethology and Evolution

MARC BEKOFF

Marc Bekoff is professor in the Department of Environmental, Population, and Organismic Biology at the University of Colorado at Boulder. He is a co-editor (with Dale Jamieson) of Interpretation and Explanation in the Study of Animal Behavior. *After a conventional training in biology, he studied the social behaviour and ecology of coyotes, wolves and dogs. He has also done fieldwork in Antarctica, on the behaviour of Adèlie penguins, and in Colorado, on evening grosbeaks. His experience as a field biologist has convinced him that we cannot study animal behaviour without investigating the animals' awareness and ways of thinking. Here he draws on that viewpoint in order to argue for the continuity of human and animal experience.*

Some Personal Reflections

On the basis of (a) common sense, (b) findings in cognitive ethology (the study of animal thinking, consciousness and mind) and (c) the notion of evolutionary continuity, a strong case can be made for admitting great apes into the community of equals. Initially, I was incredulous that such an appeal was even necessary. Next, I found it difficult to conceive that this plea could ever be denied, not only to great apes, but to most nonhuman animals. Considering, however, how many nonhumans are used by humans for anthropocentric ends, I came to see that it was my own stance that was unusual and in need of justification.

Thinking about great apes as members of the community of equals has also made me think about some aspects of what it is like to be a scientist. My early training was an instance of what Bernard Rollin calls 'the common sense of science',[1] in which science is viewed as a fact-gathering value-free activity. There was little or no overt expression of concern for the plight of nonhumans, and questions concerning morals and ethics rarely arose. When such questions did surface, they were invariably dismissed by invoking a vulgar form of utilitarianism, in which suspected costs and benefits of animal use were assessed from the human's point of view with little or no concern for the nonhuman's perspective. Often, it was simply asserted that the animals really didn't know or care what was going on. This apathy and remoteness from the animals' points of view bothered me deeply. I soon formed the opinion that ethical issues are integral and legitimate parts of science; one cannot be neutral on such matters. 'Moral privatists' who dispense with their moral and ethical obligations to nonhumans are taking a position on matters even though they are not aware of doing so.[2]

My own laboratory and field experience in ethology showed me that all behavioural research involves intervention, even 'simple observation'. I asked myself just what humans do when they study nonhumans and questioned what science was all about. I engaged in these thoughts not because I wanted to terminate my career or others' research, but because I thought it reasonable to think about behavioural biology specifically, about science in general, and also about how I spent a lot of my time. Among my scientific colleagues, my contemplation of animal welfare and questioning of science is not always well received. Recent accusations by some prominent scientists concerning the presumed intentions and characteristics of those who are interested in animal rights and animal welfare have made it essential to stress that I am not (and have never been) anti-science[3] or an anti-intellectual or Luddite.[4] Furthermore, I certainly do not want to end all research relating to animals.[5] For example, if some primates must be kept in protected areas because they were bred in captivity and it would not be in their interests to return them to the wild, or the habitat from which they came has disappeared, research that could lead to improvements in their welfare, and caused them no harm, would be permissible. Some research on animals who had already been injured, either physically or psychologically, might also be acceptable if the animals were not to suffer any further harm. In general, as far as apes are concerned, some studies would be permissible on the same basis as they would be permissible on humans incapable of giving consent: that is, where a guardian appointed to represent the best interests of the apes would give consent. While I value science, I do not worship the enterprise of science, and one does not have to be anti-science or anti-intellectual to question how science is done.[6]

Great Apes and the Community of Equals

There seem to be at least three interrelated reasons why great apes should be included in the community of equals. The first concerns the use of common sense to describe and to explain animal behaviour, the second is motivated by recent work in comparative and evolutionary cognitive ethology and the third centres on the notion of evolutionary continuity, stemming mainly from the work of Charles Darwin.

Common-sense approaches to the study of animal behaviour are useful in furthering our understanding of the behaviour of animals. The ways in which humans describe and explain the behaviour of non-humans can strongly colour views on animal welfare. Without the use of common sense and familiar anthropomorphic terms, descriptions and explanations of animal behaviour are tedious and inconvenient; they frequently lack context and content, and do not tell us very much (if anything) about what might have occurred in a given situation. Even if explanations of animal behaviour based on common sense or folk psychology are sometimes wrong (as are 'scientifically' based explanations), they can also be correct.

Data from comparative evolutionary studies in cognitive ethology – investigations into animal thinking, awareness and consciousness – also support the suggestion that great apes should be admitted to the community of equals. It has become clear that many nonhumans have rich cognitive and intentional lives[7] and also have the capacity to experience pain and to suffer; studies in cognitive ethology inform questions concerning animal cognition and animal welfare, especially when the level of development of an individual's (or species') cognitive abilities are used as a basis for moral and ethical decisions. The richness of the cognitive and intentional lives of great apes is particularly evident, as many of the essays in this book make clear.

There is also an evolutionary reason for claiming that great apes deserve membership in the community of equals. Evolutionary continuity is widely accepted by biologists, even when it is not very apparent. Ethological studies of nonhuman primates in general, and great apes in particular – species with which humans are most closely evolutionarily continuous – can certainly inform and motivate investigations into human behaviour.[8] We readily acknowledge evolutionary continuity in physiology and anatomy, and we should do so in behaviour. Some common sense is important here: can we really believe that humans are the only individuals with feelings, beliefs, desires, goals, expectations and the ability to think about things?

Those who deny that animals have beliefs or desires or propositional attitudes of different orders must offer alternatives that will be as useful for describing and explaining animal behaviour, and this they have not done. How can one deny that a great ape has some beliefs about what he or she is doing, even if the ape's beliefs are not like ours? While I do not really know that an ape expects (or has a belief) that food will be forthcoming when he or she engages in a behaviour that most humans call 'begging', I feel that this word adequately describes what he or she is doing. Likewise, when Jethro chases another ape up what I call a 'tree', it does not really matter whether Jethro knows that it is an 'ape' who ran up a 'tree'; Jethro does not need to have the concepts of 'ape' and 'tree' to engage in what we call 'ape-chasing'. I use English, the language with which I am most familiar, to convey information about Jethro to others who also understand English. I also use anthropomorphic words and phrases that seem most likely to capture the essence of what he is doing. I could describe, in great and inconvenient detail, Jethro's behaviour from anatomical and physiological perspectives, but this approach would convey little or no useful information to another person about what Jethro was doing.

Of course, it is very difficult for one to know with certainty what is going on in the minds of other individuals, human or nonhuman. Some sceptics conflate the difficulty of learning about animals' cognitive lives with the impossibility of doing so. Stephen Stich claims that because we cannot say what it is that an animal believes – because we cannot precisely ascribe content – it is fruitless to suggest that we can explain an animal's behaviour in terms of desires and beliefs.[9] More plausibly, Donald Davidson, although he is sceptical about animals having beliefs, notes that we have no general and practical alternative for explaining animal behaviour other than by attributing beliefs, desires and intentions.[10] He also holds that although language is not necessary for thought, it is difficult to imagine that there would be much thought without language.[11] None the less, Davidson does not believe that the possible absence of thought or of propositional attitudes in nonhumans means that nonhumans may be mistreated. Furthermore, demanding that nonhumans have language (as we know it) before they can have propositional attitudes, and requiring that animals have propositional attitudes before they can be granted rights or their interests receive equal consideration to those of humans, is anthropocentrically self-serving and asserts an extremely narrow view of what it is like to be a nonhuman animal.

Investigation into the cognitive skills of nonhumans frequently has surprising results, and it is essential that people who write about animal issues be cognisant of these findings. It is difficult to imagine how any coherent thoughts about moral and ethical aspects of the treatment of animals could be put forth without ethological, evolutionary and philosophical input. Clearly, ethologists must read philosophy and philosophers must not only read ethology but also watch animals.

Conclusion: The Importance of Ethological Research

Humans need to make serious attempts to look at things from a nonhuman point of view and to try to discover answers to the fascinating question of how animals interact in their own worlds and why they do so. There is no substitute for careful ethological research. While some animals seem to respond in the same way as humans to a wide variety of stimuli that are known to us to be pleasurable or painful, and arguments from analogy are often very convincing, we also know that many other animals process sensory information differently from humans and that they perform motor activities that are unlike any that humans typically perform. In these cases, arguments from analogy may fail, but this does not mean that they always fail. Furthermore, we should not conclude that any animals, human or nonhuman, cannot do something when they do not do what we expect them to do. We must be sure that the animal can perceive the necessary stimuli, is able to perform the motor activity that we think he or she should perform, and is motivated to perform this task. Furthermore, when an animal makes what we call an 'error', it may not be an error in the context in which it was made – when the animal's sensory and motor worlds are taken into account.

Adopting a common-sense approach to how we view the cognitive skills of nonhumans and their pains and suffering will make this a better world in which humans and nonhumans can live compatibly. Common-sense intuitions about pain, suffering and animal cognition should be combined with reliable empirical data, of which there are already plenty. Some claim that cognitive explanations have yet to prove their worth when compared with reductionist behaviouristic explanations of the behaviour of nonhumans.[12] Such sceptics ignore a wealth of data that demonstrate rather impressive cognitive skills in many nonhuman animals; they may also mislead those who look to cognitive ethology to provide information for structuring their ideas about animal welfare, and conclude that there is little or nothing in cognitive ethology that is convincing. What we believe about the cognitive capacities of nonhumans affects our thinking about animal welfare; different views lead us to look at animals in particular ways. Ascribing intentionality and other cognitive abilities to animals is not moot, because it has moral consequences. Common sense, findings in cognitive ethology and the idea of evolutionary continuity strongly support the present appeal for admitting great apes into the community of equals. In the future, after great apes have been granted membership, it might be wrong to ignore other species. Data from comparative and evolutionary studies in cognitive ethology and arguments based on evolutionary continuity also point to a broader view. As Degler rightly states, we must revisit Darwin and draw inspiration from 'his insistence upon the continuity between human and animal experience'.[13]

Acknowledgements – I thank Anne Bekoff, Lori Gruen, Susan Townsend, Carol Powley, Dale Jamieson, Bernard Rollin, Deborah Crowell and Robert Eaton for comments on this paper. The editors of this volume also made extremely valuable comments.

Notes

1. B. Rollin, *The Unheeded Cry: Animal Consciousness, Animal Pain and Science* (Oxford University Press, New York, 1989).
2. D. Jamieson, 'Experimenting on animals: a reconsideration', *Between the Species*, vol. 5 (1985) pp. 4–11; M. Bekoff and D. Jamieson, 'Reflective ethology, applied philosophy, and the moral status of animals', *Perspectives in Ethology*, vol. 9 (1991) pp. 1–47; M. Bekoff, 'Scientific ideology, animal consciousness, and animal protection: a principled plea for unabashed common sense', *New Ideas in Psychology*, vol. 10 (1992) pp. 79–94.
3. I.S. Bernstein, 'Breeding colonies and psychological well-being', *American Journal of Primatology*, suppl. 1 (1989) pp. 31–6.
4. Bernstein, 'Breeding colonies'; C.S. Nicoll and S.M. Russell, 'Analysis of animal rights literature reveals the underlying motives of the movement: ammunition for counter offensive by scientists', *Endocrinology*, vol. 127 (1990) pp. 985–9.
5. W.A. Mason, 'Primatology and primate well-being', *American Journal of Primatology*, vol. 22 (1990) pp. 1–4.
6. Rollin, *The Unheeded Cry*.
7. For numerous and diverse examples among which are included communication and deception, play, vigilance (antipredatory) behaviour, monitoring social relationships, the discrimination of kin and other individuals, tool use, food-caching and injury-feigning, see D.R. Griffin, *Animal Thinking* (Harvard University Press, Cambridge, 1984); Rollin, *The Unheeded Cry*; Bekoff and Jamieson, 'Reflective ethology'; C.A. Ristau (ed.), *Cognitive Ethology: The Minds of Other Animals*; D.L. Cheney and R.M. Seyfarth, *How Monkeys See the World: Inside the Mind of Another Species* (University of Chicago Press, Chicago, 1990).
8. W.G. Kinzey (ed.), *The Evolution of Human Behavior: Primate Models* (State University of New York Press, Albany, 1987); D. Bickerton, *Language and Species* (University of Chicago Press, Chicago, 1990); J. Rachels, *Created from Animals: The Moral Implications of Darwinism* (Oxford University Press, New York, 1990); C.N. Degler, *In Search of Human Nature: The Decline and Revival of Darwinism in American Social Thought* (Oxford University Press, New York, 1991); P. Lieberman,

Uniquely Human: The Evolution of Speech, Thought, and Selfless Behavior (Harvard University Press, Cambridge, MA, 1991); J.D. Loy and C.B. Peters (eds), *Understanding Behavior: What Primate Studies Tell Us About Human Behavior* (Oxford University Press, New York, 1991).

9. S. Stich, *From Folk Psychology to Cognitive Science* (MIT Press, Cambridge, MA, 1983), p. 18.

10. D. Davidson, 'Rational animals', in E. LePore and P. McLaughlin (eds), *Actions and Events: Perspectives on the Philosophy of Donald Davidson* (Basil Blackwell, New York, 1985), pp. 473–80.

11. But see L. Weiskrantz (ed.), *Thought Without Language* (Oxford University Press, New York, 1988).

12. P. Colgan, *Animal Motivation* (Chapman and Hall, New York, 1989), p. 67; M.C. Corballis, *The Lopsided Ape: Evolution of the Generative Mind* (Oxford University Press, New York, 1991), p. 24; T.A. Sebeok, 'A personal note', in M.H. Robinson and L. Tiger (eds), *Man & Beast Revisited* (Smithsonian Institution Press, Washington, DC, 1991), p. xii.

13. Degler, *In Search of Human Nature*, p. 329; see also Rachels, *Created from Animals*.

10

What's in a Classification?

R.I.M. Dunbar

R. I. M. Dunbar studied philosophy and psychology at Oxford University before completing a doctorate in psychology at the University of Bristol in 1974. Since then he has held research fellowships at the University of Cambridge and the University of Liverpool, and teaching posts at the University of Stockholm and University College, London (where he is currently professor of biological anthropology). His main research interests lie in the evolution of mammalian social systems, and the application of Darwinian principles to the study of behaviour. His recent books include Reproduction Decisions *and* Primate Social Systems. *His contribution to this project asks whether biological classifications – such as those that put us into one category and the great apes into another – are real or artificial.*

Our taxonomic classifications derive, by a rather tortuous route, from the common-sense classifications of everyday life. If you look around you, you see that organisms (and, indeed, all phenomena) can be classified into types that more or less resemble one another. These traditional classifications are, in some genuine sense, 'natural'. They are also functional, in that classificatory systems serve a purpose for the individuals that make them. We classify phenomena in order to reduce the information overload that would otherwise occur, but exactly how we classify them is largely a matter of convenience. For this reason alone, different cultures are likely to end up classifying phenomena in slightly different ways.

Since the rise of biology as a formal discipline founded in evolutionary theory, the classification systems developed by biologists in the latter part of the nineteenth century came to serve a more specific purpose: they were intended to mirror the branching history of evolution. Unluckily, the way in which the classifications were carried out was often both circular and rather subjective. You more or less decided in advance who was related to whom, and then looked for anatomical evidence to support your hunch.

This unsatisfactory state of affairs was gradually resolved during the second half of the twentieth century with the rise of what is now known as cladistics. The cladistic approach essentially asks us to classify organisms on the basis of straightforward anatomical similarity. Subsequently, you can, if you wish, infer something about their evolutionary history from the patterning of the relationships between these clusters. But it is essential that inferences about evolutionary history come after classifying, and not before.

Whichever way you look at it, chimpanzees, gorillas and orang-utans turn out to be our closest living relatives. They share many anatomical features with us that mark us all out as belonging to the same classificatory group, the hominoids or human-like animals. These features include such things as a peculiar pattern to the molar teeth, a flattened chest, the lack of a tail and so on. It is also likely that they share a relatively recent historical ancestry with us. During the last decade, advances in molecular genetics have revealed that the genetic similarities between humans and the African apes are very close indeed: in genetic terms, the differences are trivial, especially those between ourselves and the two species of chimpanzees. Not surprisingly, perhaps, it's particularly easy to empathise with the African apes: they seem somehow to be more human-like in their behaviour as well as in their appearance.

In addition, the chimpanzees share with us a number of psychological characteristics that have not been found in other species. One of these is the ability to engage in pretend play; another is to be able to see the world from another individual's point of view. Some human beings (namely autistic individuals) lack both these abilities, yet we are happy to treat them (quite rightly, of course!) as human. How much more deserving then must be the chimpanzees case for equal treatment!

The biological reality is that all classifications are artificial. They force a certain order on to the rather chaotic mess of the natural world. Species, as we describe them, are matters of convenience rather than biological reality. The real world consists only of individuals who are more or less closely related to each other by virtue of descent from one or more common ancestors. It may not be unreasonable to argue that the only grouping that has any biological validity is the local population of individuals who breed with each other. By contrast, classifications of species based solely on physical appearance can be misleading. Sometimes, relatively rapid changes in appearance take place as populations expand and disperse over a wide geographical area: the result is organisms that look very different (and may get classified as different species) even though they share a recent common ancestor. In other cases, rather little change in physical appearance may take place over a very long period of time: the result may be populations of organisms that look very similar, but whose common ancestor lived a very long time ago. Appearances can be deceptive, and are not always a reliable guide to pedigrees.

The domestic dog provides us with a familiar example. The wild Australian dingo looks rather similar to the native dogs from many parts of the Old World, yet the last common ancestor that it shares with the rest of the domestic dogs of the world must have lived some tens of thousands of years ago. In contrast, the many breeds of pedigree dogs look very different not just from each other, but also from the dingo. Who would guess that the dachshund and the Alsatian share a common ancestor that lived no more than a few hundred years ago? Who would, in all seriousness, have thought of classifying the Pekinese and the greyhound in the same species? By all the rules of taxonomic classification, we ought to class them in different species. The only reason we do not is that we know that they have been deliberately bred to look this way within recent historical times. And, of course, genetically they are all but identical.

When the histories of two types of animals are unknown because they are buried in the fossil record, we lack the privileged information about pedigrees that would allow us to specify degrees of relatedness. Instead, we must rely solely on physical appearance.

Now, human beings the world over spend a lot of their time classifying the organisms that share the planet with them. Much of that effort is devoted to distinguishing between 'them' and 'us'. We do it not just with respect to other forms of life ('It's bad to kill people, but OK to kill mosquitoes'), but also with respect to other humans. The words that most 'primitive' tribes use to refer to themselves invariably translate as 'human' or 'people'. Members of other tribes are commonly classified as 'not-human', as indistinguishable from a number of other nonhuman creatures (usually other mammals). Supposedly 'civilised' peoples like the ancient Greeks and Victorian British may not have used the same terms, but their attitudes towards subject peoples was often virtually the same. At least one likely reason for this is that it allows us to define the categories of organisms that we can legitimately exploit. (When I say 'legitimately' here, I mean simply those that we can exploit without having to feel guilty.) There are probably very good biological reasons why humans (and, for all we know, other animals) behave in this way. The point simply underlines the fact that the use we make of classifications is largely a moral and not a scientific issue.

The biological reality is that the great apes are just populations of animals that differ only slightly more in their degree of genetic relatedness to you and me than do other populations of humans living elsewhere in the world. They just look a bit different to those other populations that we commonly call 'human', but not all that different, and by no means as different as, say, spiders do.

In many ways, our perceptions of similarities and differences in these cases probably reflects the extraordinary similarity in appearance that all living humans show. Humans are much more similar to each other, both in appearance and in genetic make-up, than almost any other group of animals is. This similarity reflects our relatively recent origin as a species: the last common female ancestor of all living humans (the so-called 'mitochondrial Eve') probably lived as recently as 150,000 years ago — a mere 5,000 human generations ago! Indeed, so recently did our common ancestor live, that all humans alive today everywhere in the world are much more closely related to each other than either all the chimpanzees are to each other or all the gorillas are to each other.

This remarkable similarity between humans may help to explain why we draw the line so tightly around ourselves. In trying to differentiate between 'them and 'us', we observe that the differences in appearance between members of our own family and other humans from all parts of the world is relatively small compared with the apparent gulf between ourselves and those species that seem to be most similar to us (namely, chimpanzees and gorillas). After all, there are some pretty dramatic differences: we walk upright, build cities, compose and perform music, send rockets to the Moon, and so on.

But the apparent size of the gulf separating us from our nearest relatives is largely a consequence of the absence of any intermediate species alive today. (And I do not mean 'missing links' that represent an intermediate stage through which chimpanzees passed on their way to becoming humans; I mean species that have evolved to different stages from the same common ancestor as ourselves and the chimpanzees.) It is not that there have never been any intermediate species. Such species have existed in plenty: their remains are to be found in the fossil-bearing sites of eastern and southern Africa. Rather, our problem is that they are all extinct, so we have no way of knowing at first hand the degree to which their behaviour was like ours. If any of these species were alive today, the gulf between ourselves and the chimpanzees would almost certainly seem a lot less dramatic than it does at present. And that, in turn, would cause us to think again about how we classify chimpanzees and gorillas in relation to ourselves.

It might well cause us to classify ourselves as just another species of chimpanzee rather than in an entirely separate group of our own.

11

Apes and the Idea of Kindred

STEPHEN R.L. CLARK

Stephen R.L. Clark is professor of philosophy at Liverpool University. His publications include: Aristotle's Man, The Moral Status of Animals, The Nature of the Beast, From Athens to Jerusalem, The Mysteries of Religion, Civil Peace and Sacred Order, A Parliament of Souls, God's World and the Great Awakening; *his next book will be called* How to Think about the Earth. *He has given Gifford lectures (Glasgow), Stanton lectures (Cambridge), Wilde lectures (Oxford), Scott-Holland lectures (Liverpool) and is to give Read Tuckwell lectures at Bristol in 1994. He is married, with three children and four cats. In the essay that follows, Clark asks: how do we decide where to draw the boundaries of our own species? Is it in some way inevitable or natural that we should consider 'our kind' to be all, and only, those we now consider to be human beings? Or might we have drawn the boundaries of our species differently?*

How Were the Apes Demoted?

There was serious debate in the seventeenth and eighteenth centuries as to the precise limits of humankind. Monboddo, in particular, concluded that such apes as the orang-utan and chimpanzee (counted together as the Ouran Outang) were of our kind, a notion satirised by Thomas Love Peacock in *Melincourt*, in the person of Sir Oran Haut-ton, a parliamentary candidate for the rotten borough of Onevote.[1]

They are exactly of the human form, walking erect, not upon all-four. . . . They use sticks for weapons; they live in society; they make huts of branches of trees, and they carry off negro girls, whom they make slaves of and use both for work and pleasure. . . . But though from the particulars mentioned it appears certain that they are of our species, and though they have made some progress in the arts of life, they have not come to the lengths of language.[2]

Monboddo was gravely misinformed in some respects, and engagingly open in his judgement that our species-nature was shown chiefly in war, rape and domination rather than, as tradition said, in the use of language. He guessed right, though perhaps for not entirely happy reasons, that 'if ever men were in that state which [he] call[ed] natural, it must have been in such a country and climate as Africa'.[3] Maybe he guessed wrong about our species-nature. His inclusion of apes within 'our kind' is matched by those of his contemporaries who excluded Hottentots (like Voltaire[4]). Those who insisted, with J.G. Herder, that 'neither the *Pongo* [probably the chimpanzee] nor the *Longimanus* [the gibbon] is your brother; but truly the American [that is, the Amerindian!] and the Negro are',[5] now occupy the scientific and the ethical high ground. Any attempt to re-open the question is bound to seem offensive, especially if it is conjoined with the somewhat salacious details enjoyed by earlier anthropologists and explorers. I share with liberal critics a suspicion that supposedly 'objective' examinations of, say, the brains of 'Australids' (that is, native Australians), orang-utans and 'Europids'[6] are profoundly racist in their motivation and execution. But there really are important questions here. The story of the exclusion of such apes from 'our' kind requires an examination of the relations between folk taxonomy (which is strongly evaluative) and scientific taxonomy (by which biological taxa are defined genealogically).[7] What follows is a beginning.

Species-natures and the Nature of Species

Either we are simply natural products of evolutionary processes or we are not. In this part of the chapter, I explore the former hypothesis. My conclusions are very much like those of Richard Dawkins in Chapter 7: if we are products of evolutionary processes, then any objective judge would be likely to count us together with the other apes (just as we think ants or dolphins or finches are of single kinds even though there may be many (strict) species of ant, finch or dolphin). This is not to say that all such kinds display a single nature. Chimpanzees and pygmy chimpanzees and gorillas and orang-utans and humans are different in many ways. But it does not follow that they are not of the same biological kind. To explain this in more detail than Dawkins attempts, and to show why it is that moralists have neglected this, requires careful attention. Nothing that I say in this section is really original, but it is sufficiently unfamiliar and unwelcome to sound both strange and difficult. Really, it is neither.

One of the points on which philosophers have yet to agree with biologists is that there *are* no natural biological kinds in the sense once intended. 'Natural kinds', so-called, are sets of creatures with a shared, distinctive nature, but biological taxa, including species, are not so defined.[8] Even if all members of such a taxon happen to have shared, distinctive properties that is not why they are its members. Cows are not mammals because they feed their young on milk; bovine mothers feed their young on milk because they are mammals. Being a mammal is being genealogically linked with a complex individual, the order Mammalia, such that its members are more closely related to each other than to members of any other order. This is not to say that they more closely resemble each other. The order's members, or those now judged to have been its members long ago, were not always more closely related to present-day mammals than to their non-mammalian contemporaries. Even now, there may be mammals that *look* more like non-mammals than they look like any other existing mammals. There might even be mammals whose parturient females do not secrete milk, just as there might be birds without wings or feathers. They are not therefore 'imperfect mammals', though such phrases are not wholly unhelpful. When Aristotle identified seals, for example, as 'deformed quadrupeds', he was partly right – though any implication that seals are therefore not what they should be must be resisted (Aristotle also suggested that women were deformed men!).[9] Whereas philosophers still tend to believe that there are 'typical' members of a taxon, and to be as eager as Aristotle was to identify defect or anomaly, modern biologists think that cheetahs are as obviously cats, Down's Syndrome children as obviously human, as any 'type-specimen'. Either might have *been* the type-specimen of the relevant taxon, because the biological type of a taxon is simply the specimen (however unusual it eventually turns out to be) that serves as the referential tie for that particular taxon.

Biological taxa are individuals.[10] That may seem thoroughly mistaken: surely taxa are sets of individuals who more or less 'resemble' each other? Quine stated that a biological kind is the set of all things 'to which [the paradigm] *a* is more similar than *a* is to [the foil] *b*'.[11] There may be such sets, but they are not the same as taxa. *Drosophila pseudoobscura* and *Drosophila persimilis* are sibling species, indistinguishible to naive observers, but certainly distinct (because their members do not successfully breed together). If one such species vanished from the world, but there later appeared creatures indistinguishable even to an expert eye (but having a different ancestry) the older taxon would not have reappeared. The dodo, once extinct, is gone forever, because 'dodos' are not just those creatures that look more or less like the pictures, nor even those creatures whose DNA looks more or less like that of the old birds (if we could discover this). There is no need for members of a given taxon to resemble each other more than any of them do members of another taxon. There is no need for them even to share any particular genes which are not shared with creatures of another kind. So being a member of that taxon is not a matter of instantiating any non-historical property, whether obvious or hidden away. The taxon is an individual, and ordinary individuals are parts of it, segments of a lineage. 'Academic classification extends to classes, which it divides according to resemblances while natural classification divides according to relationships, by taking reproduction into account.'[12]

Where two or more species emerge within a previously existing taxon that till then had only been *one* species, it is customary to give both of them new names, even if one such species is indistinguishable to us and to its members from the older species. If the earlier species merely develops, as a single population of interbreeding individuals, taxonomic practice varies: some will judge that the differences are such that if the species had been contemporaneous they would not form a single breeding population; others that there is no clear break between the old and new, which may as well be counted as one species. *Homo erectus*, *Homo habilis* and *Homo sapiens* can be one species or three. Traced back through time, of course, the difference between species will often seem quite arbitrary, even when there are two or more daughter species to consider. Why, after all, should x be a different species from y merely because there is another species, distinct from x but equally descended from y? If that other species had not been discovered, or had been extinguished before it was truly established, x would be uncontroversially the same species as y. Even at one time there are populations which reveal the transience of species: the different varieties of herring gull described by Richard Dawkins in Chapter 7 are one species in the sense that genes from one variety can spread, by degrees, to any other, but they are two or more species if particular varieties are paired. This phenomenon is, in Kant's terminology, a *Realgattung*, a historical collection of interbreeding populations, and in more modern terms a *Formenkreis*, or ring species. It is probable that humankind, historically, is such a *Realgattung*; it may even be that it still is, that there are particular varieties within the species that would be judged different species if the intervening varieties were lost.[13] As Dawkins points out, humans and chimpanzees are judged to be of different species precisely because the intervening varieties are indeed extinct.

Once synchronic species barriers are established, the flow of genes will be restricted: that is what species barriers are, and that is why lions and tigers are of different species. But their ancestors were not, and genes flowed equally from the urcat to lion and tiger populations. It is because there are, by hypothesis, no diachronic species barriers, that some have reckoned that palaeospecies (like *Homo habilis*) are only metaphorically species at all. Nor can we always claim that the barriers against interbreeding were established by some other general change in the character and conduct of ur-lions and ur-tigers. More likely the barriers were established, by distance or mountain or river, for reasons having nothing to do with any original characters, and the other general differences accumulated since. We cannot even be sure that what had seemed like barriers against interbreeding are always more than accidents of preferences or opportunity: maybe lions and tigers *would* interbreed successfully often enough to identify them as a ring species if enough of them had the chance. There is good reason, after all, to think that domestic dogs, wolves and coyotes are really all one, variegated species such that not all its varieties will willingly interbreed. Lose all the other dogs, and wolfhounds and chihuahuas would be unlikely conspecifics – far more so than Voltaire reckoned 'Hottentots, Negroes and Portuguese'.[14]

But surely human beings are all one species in a more important sense than this? Maybe other grades of taxon, family or order or phylum, are merely genealogically united. Maybe there are taxa that look like species but are really not. Species, real species, are, precisely, special. Don't members of a single species share a nature? Don't human beings? Isn't that the axiom on which humanism and the United Nations charter depend? If modern biological theory suggests that there need be no shared natures, no perfect types, and even that the purity of species must be questioned, so much the worse for biological theory. The Negro and the American *must* be our brothers and sisters, and therefore must be like us. It is insufferable to suggest that there are real varieties of humankind that might not willingly interbreed. Still less sufferable to imply that *Pan, Pongo, Gorilla* and *Homo*[15] might perhaps have been, or still may be, a ring species. What varieties of *Homo* might breed, or once have bred, with *Pan*? Isn't there a submerged, and prurient, racism at work here? 'Ape' is an easy, racist insult. Those whites who use it should perhaps be reminded that most commentators, worldwide, will suspect the smelly, hairy Europids (which is, 'the whites') of being 'closest' to the ape. Do we need to abandon biological science to avoid being racists?

Racism, on the contrary, is the natural expression of misplaced essentialism, the belief that groups embody different natures. Or else it is an early version of the barriers against interbreeding that establish distinct species. New species is but old race writ large. There need be no antagonism between such newborn species: the consequence may actually be a lessening of competition, because the species eventually graze in different places and on different things. There may also be a reduction in adaptive variation, since one newly distinct lineage has lost the input from the other. There may be many good reasons for us not to allow the emergence of barriers between human varieties, and maybe the other breeds of hominoid are now sufficiently distant (as once they were not) from 'ours' as to make all present interbreeding doubtful. But we do not *know* this to be true. I do not recommend the experiment – but mostly because the fate of any such cross-breed as that imagined by Dawkins would probably be to serve as laboratory material. As long as misplaced essentialism rules, we will suppose that cross-breeds do not really share our nature, that they are throwbacks to a pre-human kind not 'of our kind'. The truth is otherwise.

A further difficulty for moralists is the rejection of norms in nature. If there is no one way of life and character which best suits all or most members of a particular kind, such that we may detect deformity, disease or deviance by comparison with that ideal type, can there be 'a good human life'? Can we truthfully suggest that battery chickens are deprived by being denied 'the' life that chickens would live 'in nature'? If species are only genealogical groups, such that members need not especially resemble each other, we have no right to suppose that there is one way only (however vaguely defined) for any particular species. The limits of variation will be empirically discoverable: what a kind can adapt to will be shown by what as a matter of fact it does, and there will be nothing normative about its 'natural' life. Lineages evolve to make and fit their environments, or else are extinguished. I am a member of the Clark family: but not because I resemble other Clarks, nor yet because there is a way that Clarks will naturally live that is unlike the way that others live. Even if Clarks were more inbred than they are (and so approximated the condition of a species) they need not always resemble each other. There might be atavisms, sports, changelings or disabled Clarks, but they would all be Clarks, and such variations would not be failures: on the contrary, they would contain the Clarks' hope of posterity. Variation is not a dysfunction of sexual reproduction (even if animal breeders are annoyed if the line they are concerned about does not 'breed true'): it is what sex is for.

This may seem old news. After all, moral philosophers have insisted for most of this century that no natural facts are norms, that 'natural' is not necessarily a term of praise. They have even insisted that human beings are special because they have no single species-nature. The claim is flawed: partly because nothing has such a nature, and partly because the claim exactly identifies a nature shared, at least in potency, by every human being.[16] As Aristotle said, we are creatures whose life is one of acting out decisions. Aristotle was less essentialist than his heirs, because he never identified that 'we' straightforwardly with a species.

Not all those born into our species will be capable of their own actions. In some the capacity for choice, or even for understanding, is missing from the start. The good life for us is the life well-lived by those who ask such questions. The very same moralists who have emphasised our freedom from natural constraint actually think but poorly of those who do not, or even cannot, make their 'own' decisions, and constantly deplore attempts to make divisions within the species, which we ought (by their account) to be entirely free to make. So once again the primary danger is not from the biologically grounded notion of kinds, but from our habitual confusion of species and natural kind. We can find out a lot about what individual creatures need and like, what sort of lives they can arrange to live together. We do not need to think that there are goals that only and all conspecifics share.

A fully Aristotelian ethic can accommodate the remarks on species-natures that modern biologists endorse. 'We' means only those engaged, or potentially engaged, in this sort of conversation. 'We' are probably all human, as being members of the *Realgattung* of humankind. But not all our conspecifics need be self-motivating or rational in any way. 'Maybe all triangles must have three angles, but not all reptiles must have a three-chambered heart, though in point of fact they might'.[17] By the same token, Monboddo was right to think that not all human beings *must* be able to speak. And creatures who are not now of our species, though their ancestors once were, may share as much with us as any disabled human. The thought is a dangerous one, no doubt. We are not long removed from moralists who deployed Aristotle's ethics to suggest that Amerindians were natural slaves, owed no respect as real images of God. Any suggestion that not all our conspecifics share our nature is heard as licence to oppress and kill. But there is no such licence, nor any proper argument from neo-Aristotelian premisses to ignore the ties of kindred.

Who Is My Sister?

A fully modernised neo-Aristotelianism is probably inadequate, for reasons that I shall address below. But it may serve us for a moment. UNESCO's declaration that 'all men [sic] belong to the same species' was clearly intended as a moral commitment to the thesis that all human beings have very similar needs, and that those needs should be met by the global community, acting through the various national authorities that are the best we can yet manage (however obviously imperfect they are). It was certainly a necessary commitment, in the face of those who had sought to divide the species against itself, and create a new, predatory species on the ruins of the old. As a moral programme, some will say, there was no need to give reasons, or to found our commitment in any agreed facts. That we are all one species was never really a biological claim, as might be a similar assertion about dogs, wolves, dingoes and coyotes (but not marsupial wolves). Nor would its authors, we must suppose, have been really alarmed to find that humankind was rather a *Realgattung* that might easily become two or more species. They might not even have been moved by the discovery (if such there could be) that not all our conspecifics possessed rational souls, or were capable of reasoned action. Moral commitments need not and cannot, so we have often been told, rest on any non-moral dictum.

This is not the time to explore that particular error, except to say that it *is* an error. If non-moral dicta are ones that can coherently be joined both to one moral dictum and its opposite, then it is, of course, impossible to demonstrate either one of those moral dicta from a dictum that, by definition, does not have moral implications. *P* cannot strictly imply *Q* if (*P* & *Q*) and (*P* & −*Q*) both make sense. It does not follow either that there are such 'non-moral dicta', or that strict demonstration is the only rational form of argument, or that there are no ordinarily factual dicta with moral implications. At the very least, ordinary moral argument is possible, and most of us expect to give some reasons, even if not demonstrative reasons, for the causes we endorse. UNESCO's declaration was not intended as an arbitrary judgement, a dictat of the world's new rulers: its authors obviously thought that those who disputed it were *wrong*.

They thought, that is, that our conspecificity should make a difference to the extent and nature of our obligations. Whereas what is now called 'racism' claims the right to treat human beings of other races less favourably than the racist's own, UNESCO's demand was that no differences of race, sex, age, intellect, capacity or creed should license what would otherwise be obvious injustice. It may be that one historical explanation of the slogan's popularity in the West, in addition to the shocked discovery of what racist jibes about 'backward races' had led to in the West, was the converse discovery that, for example, the Japanese so heartily despised the smelly, hairy Europids they captured. We all began to realise how vulnerable we were.

The natural conclusion has been that species differences do license such injustice, perhaps because such differences are real and predict-able, and relevant to the nature of the putative injustice. Those reasons are not wholly wrong, but of course they hardly touch the real point: some of our conspecifics would not be injured by acts that injure us, just as some creatures not of our species would be injured by those acts. If what matters is only the quality of the putative injury, then there will be many occasions when, if we ought not to injure those capable of being injured, we ought not to injure those outside our species, and may do to our conspecifics what, in their case, will be no injury. That we are conspecifics plays no central role in the argument. Nor are any of the merely rationalist arguments very successful: respect for humankind's unity is not well represented by respect for rational autonomy, since not all human beings are thus rational. If UNESCO wished to oppose the Nazi project (as of course it did) it could hardly do so by endorsing the Nazi preoccupation with such forms of human living as they deemed rational. The object of the declaration was to oppose the extermination of the 'mentally unfit' or 'the backward races', and not merely to dispute the Nazis' identifications, as though their error was only a case of mistaken identity.

So it seems likely that conspecificity was really offered as a moral fact, a fact that ought to influence our actions and omissions. 'We be of one blood, you and I' was the slogan taught to Mowgli for his safety's sake.[18] In that fictional case the slogan is a sort of magic, which compels courtesy even if it is literally false; or else it is a promise to live by the same law as is invoked, the rule of mercy. But part of its force, and the reason for the metaphor, is that we do care for our kindred. Because we share our ancestors and may yet share descendants, because we live under the same sky, are nourished by the same foods, share the same diseases even, we are part, and know ourselves a part, of one individual lineage. What we do to others of our kind we do to ourselves, because we are all one, variegated kind. It is not that we are all or most of us *alike*. Our pleasure in each other is that we are *different*, and yet the same.

The moral truth that lies behind the error that others have called 'speciesism' (and justly rebuked) is that we both do and should treat those 'of our kind' better. But now recall the remarks I have already made about the actual nature of a species. We retain the word as physicists retain the word 'atomic'. But modern atoms are not *a-tomic* (which is indivisible), and modern species are not specific (any more than Aristotle's were[19]). Species kinship rests on relationship, and not resemblance, although there will be various similarities to reckon with within and without the kind. UNESCO's sloganeers did not recognise that *Pan*, *Pongo* and *Gorilla* were our sisters, any more than the writers of the American Declaration of Independence fully realised what they had committed themselves to by saying that all *men* were equal. They and their successors could have insisted that no mention was made here of women, or that the obvious intention at the time was not to include Negroes (since the passages denouncing Britain's involvement in the slave trade were, of set purpose, omitted from the final document). Instead, the real implications were allowed to emerge. All those of one kind with us begin as equals: we are, each one of us, a part of one long, variegated lineage, sharing enough of our habits, gestures and abilities to reveal our common source.

The real danger to a decent humanism (that is, to the rule of law, the rejection of oppression and genocide) is not from those who emphasise our kinship with the other apes, but from those who rest the demands of humanism only on resemblance. Resemblances are easily denied or altered; historical relationships are not. Not all our kindred are adept at language of the familiar, human kind; not even all our ordinary conspecifics are. It is enough that we are apes together, and know from the inside what it is usually like to be a versatile, manipulative primate with a sense of family and friendship. The American and Negro are indeed my sisters and my brothers; so also are the other apes who form, with us, the great individual, *Hominoidea* (which is the greater humankind).

What then is the good life for *Hominoidea*?[20] Hominoids live well when they can gather in friendship, in groups small enough that their individual status can be recognised and large enough that they can find congenial companions. They live well when they are fairly secure from arbitrary arrest and murder, with some opportunity to innovate and to explore. Who can secure this for them? Plainly, at present, only those with power and foresight enough to face and solve the problem. There is a risk that any such creative minority will conceive itself to be an intellectual and political elite, and run things only for its immediate interest. It is certainly a good thing that recent elites have been constrained at least to pay lip service to the slogan that all human beings are 'equal' and all owed a like respect which they cannot enforce. Demanding a similar respect for hominoids (including, please remember, us) may place too great a strain on intellectuals still in the grip of fantasy. With care, however, we may risk the attempt.

The Spiritual Form of Humanity

I said before that either we are simply natural products of evolutionary processes or we are not. If we are, then it seems clear that there are no rigid boundaries between species groups, that species, and other taxa, are quite real, but only as *Realgattungen*. There is a real difficulty, however, in believing this, despite the efforts made by other contributors to this very volume to expound a fully naturalised epistemology. The argument, which is a powerful one even if it has not convinced all theorists, runs as follows. If we are the products of evolutionary processes, then we have no good ground for thinking that our thoughts are anything but none-too-harmful fantasies. As Nietzsche saw, we must presume that we have evolved as the descendants of creatures who could ignore a lot, who could live out their fantasies. There is nothing in evolutionary epistemology to give us reason to expect that we would care about the abstract truth, or ever be able to obtain it. If the theory is correct, we have no reason to think that we could find out any correct theories, beyond (at best) such truths or falsehoods as we need to obtain the next meal or avoid being one. And so we have no reason to suppose that any theory that we have devised is really true, including the current theory of evolution. Only if the divine reason is somehow present in us can we expect that we could find out truths, or trust our moral instincts. That, after all, was what Enlightenment thinkers thought, borrowing a Platonic doctrine about the powers of reason that does not fit the neo-Aristotelian framework I have so far described.[21]

This alternative picture – that evolutionary theory does not leave room for the kind of being we have to think we are (namely truth-seeking and would-be moral images of a divine reason) – is what has often lain behind attempts to insist upon a radical disjunction between apes and people. But there is a better answer. Plato, after all, denied that it was sensible to contrast human and nonhuman things, creatures of our specific kind and all others. We might as well divide the universe into cranes and noncranes.[22] By his account (or at least the account developed from his writings), there are indeed real natures, but they are not identical with the things that partly remind us of them. Even we ourselves are not wholly identical with the Form of Humanity, though we are called to serve it. The Form of Humanity is the divine reason, and we are indeed more *human*, in this sense, insofar as we think and do as the divine reason requires. The true image of humanity, for us, is the saint or perfect sage.

How are such forms related to the *Realgattungen* I have described before? Simply enough: one lineage after another has tuned in (as it were) to the Form of Life-at-Sea, and so produced the lesser kinds of shark, mackerel, plesiosaur, whale, dugong, seal. Life-in-Society has found its images among the ants, termites, bees and mammals. Maybe we can bring the Form more clearly into temporal existence, but we can hardly do so by denying all its influence on beings outside our immediate kin, as if it were our possession. If saints are the ones who best embody Humanity (as the Platonic tradition would suggest), then we will do the best we can by imitating them. From which it follows that we should respect those other 'apes', our kindred.

If there are no natural kinds but only *Realgattungen*, then it is reasonable to think of ourselves as parts of *Hominoidea*, 'greater humankind'. If there are such kinds as Platonists imagine, and 'we' here now are partial imitations of the Form of Humanity, let us imitate it better by being humane. Such Forms require no special ancestry, nor can we boast of being children of Adam, as if the Creator could not raise up new children of the spirit from the dead stones. That spirit, if it is so truly universal as to contain the truth of things (which is the condition of our finding truth), must also be present everywhere. If we should respect Humanity in ourselves and others we should, by the same token, respect the other creatures that reflect that Form in however tarnished a mirror. If we are apes, let us be apes together. If we are 'apes' (as aping the Divine), let us acknowledge what our duty is as would-be saints and give the courtesy we owe to those from among whom we sprang. Either we evolved along with them, by the processes described elsewhere, or else we evolved, in part, to imitate a Divine Humanity. Neither theory licenses a radical disjunction between ourselves and other apes. Either may give us reason to esteem and serve the greater humankind.

Notes

1. T.L. Peacock, *Melincourt* (1817) especially ch. 6.
2. Lord J.B. Monboddo, *Of the Origin and Progress of Language* (Kincaid & Creech, Edinburgh, 1773–92), cited in J. Baker, *Race* (Oxford University Press, London, 1974), p. 23.
3. Monboddo, *Origin*, book ii, ch. 5, cited in Peacock, *Melincourt*, ch. 6.
4. See Baker, *Race*, pp. 19f.
5. Cited in Baker, *Race*, p. 22.
6. See Baker, *Race*, pp. 292ff., after E. Smith (1904).
7. S.R.L. Clark, 'Is humanity a natural kind?', in Tim Ingold (ed.), *What is an Animal?* (Unwin Hyman, London, 1988), pp. 17–34.
8. See E. Sober, 'Evolution, population thinking and essentialism', *Philosophy of Science*, vol. 47 (1980) pp. 350–83.
9. On which see S.R.L. Clark, *Aristotle's Man* (Clarendon Press, Oxford, 1975), ch. 2.2; and S.R.L. Clark, 'Aristotle's woman', *History of Political Thought*, vol. 3 (1982) pp. 177–91.
10. D. Hull, 'Are species really individuals?', *Systematic Zoology*, vol. 25 (1974) pp. 178–91.
11. W.V. Quine, 'Natural kinds', in N. Rescher (ed.), *Essays in Honour of C.G. Hempel* (D. Reidel, Dordrecht, 1969), p. 9.
12. I. Kant, *Gesammelte Schrifften*, vol. 2, pp. 427–33, cited in Baker, *Race*, p. 81.
13. For further details of modern taxonomic practice see E. Mayr, *Principles of Systematic Zoology* (McGraw-Hill, New York, 1969); C. Jeffrey, *Biological Nomenclature* (Edward Arnold, Cambridge, 1973); C.N. Slobodchikoff (ed.), *Concepts of Species* (Dowden, Hutchinson & Ross, Stroudsberg, PA, 1976).
14. *Questions sur l'encyclopédie*, cited in Baker, *Race*, p. 20.

15. All these are genera, with several species usually included in them (as *Pan satyrus paniscus*, the pygmy chimpanzee). Linnaeus identified the chimpanzee instead as *Homo troglodytes*. Nowadays, the level of the taxon is determined by professional judgement having to do with presumed ancestry and relative degree of relatedness.

16. See S.R.L. Clark, 'Slaves and citizens', *Philosophy*, vol. 60 (1985) pp. 27–46; and S.R.L. Clark, 'Animals, ecosystems and the liberal ethic', *Monist*, vol. 70 (1987) pp. 114–33.

17. D. Hull, *Philosophy of Biological Science* (Prentice-Hall, Englewood Cliffs, NJ, 1974), p. 79.

18. R. Kipling, *The Two Jungle Books* (Macmillan, London, 1910), p. 49.

19. Since scientists generally learn their Aristotle solely from Enlightenment anti-scholasticism, their picture of Aristotle is unrecognisable to any competent Aristotelian scholar; see D.M. Balme, 'Aristotle's biology was not essentialist', *Archiv für Geschichte der Philosophie*, vol. 62 (1980) pp. 1–12.

20. This is technically a superfamily, including *Pongidae* and *Hominidae* as families. The pongids include gibbons and siamangs as well as the great apes and now extinct varieties such as *Ramapithecus*. *Hominidae* includes *Australopithecus*, *Pithecanthropus* and the various, mostly extinct, species and palaeospecies of *Homo*. These classifications, remember, are still guesses about the likeliest historical relationships. *Pithecanthropus* is classed by some in *Homo*.

21. See further S.R.L. Clark, *God's World and the Great Awakening* (Clarendon Press, Oxford, 1991).

22. Plato, *Politicus*, 263d.

12

Ambiguous Apes

RAYMOND CORBEY

Raymond Corbey, a philosopher and anthropologist, teaches in the Department of Philosophy of Tilburg University, The Netherlands. Most of his research and publications concern the history and backgrounds of changing and conflicting interpretations of human nature, human origins and our position vis-à-vis animals. He has studied the views taken both in Western and in non-Western cultures on these questions, and has examined not only the work of philosophers and anthropologists, but also other cultural contexts, such as museums, world fairs and missionary discourse. Here he looks at the way in which we distinguish ourselves from the other great apes, while at the same time recognising ourselves in them.

Suppose some part of the tropics were inhabited by several highly developed societies of apes far more intelligent than humans, having a form of state and government, several kinds of industry, sophisticated technology and all institutions that come with complex societies, such as hospitals and universities, zoos and museums. Let us assume that on the periphery of these societies of apes, whose population is numbered in the millions, a few thousand humans, close biological relatives of these apes, are still to be found. The apes look upon these humans, an endangered species, as dull, uncivilised, and indeed unapish, lower beings. Until recently, relatively well-off apes used to go out and hunt humans, bringing back their hands and heads as trophies to decorate their interiors. Museums sent out their staff to shoot humans and bring back their bodies to be studied, stuffed and put on exhibit. This is not done any more now, but humans are still be to be seen in zoos, and in circuses performing tricks of all kinds, much to the amusement of ape children.

There are many universities in these ape lands, and at all of them humans are bred and studied. One ape professor became famous with experiments on emotional deprivation. He isolated a large number of human babies right after birth, and let some of them grow up in complete solitude, others with several kinds of puppets; some of the puppets provided milk, but no physical warmth, others both milk and warmth, but no movement, and so on. These experiments with human specimens provided important insights into the psychological development of young apes. Many medical departments breed humans or buy freshly caught ones to test new medicines on, especially in connection with lethal viral infections. Many laboratory humans suffer and die each year, but the experimenters claim that by these methods the lives of many apes may be saved. And what, as several professors of theology, philosophy and ethics at these universities in ape land are ready to explain, counts for more, the life of some low, unintelligent, brutish mere human, or that of a real ape, living up proudly to the high moral standard which is implied in her or his ultimate origin in Apasia, that is, the abstract, transcendent principle of reality, which created the apes, and only them, after its own image?

Now let's vary this imaginary world somewhat, and replace the intelligent apes by, say, a complex society constituted by friendly, highly civilised, carnivorous or omnivorous pigs. In this society of intelligent and altruistic pigs, tens of thousands of humans are bred and killed every year as food. They live their short, joyless lives crammed together in small, dark cells, being fattened and injected with hormones before they are processed into sausages and burgers. And the friendly theologians in this society would be among the first to point out that this is the natural order of things, that it is how the Creator – which, they add, is of course just a figurative way of speaking – wanted it to be; that pigs themselves are never to be killed or eaten because they are real persons, referred to as 'she' or 'he', and not, as in the case of brute humans, as 'it'.

What do these thought experiments teach us? To begin with, they create an estrangement which serves as an eye-opener. They confront us with the fact that the way in which we deal with apes, pigs and other animals in our society, in our thinking as well as in our daily practice, is not the only possible way. They open our eyes, blindfolded by custom as they are, to the accidental nature, and indeed the brutality, of a cultural order we perceive as 'natural', an order founded on the pivotal, apparently unquestionable idea of human moral and ontological supremacy over the rest of nature. More specifically, this topsy-turvy world raises the question *why* we, or at least most of us, accept the ways in which we usually think and feel about, and behave *vis à vis* animals as unproblematic. I shall start with some reflections on this matter and on our treatment of animals in general; afterwards I shall deal with traditional views of apes in our culture and show how and why these are beginning to change. I shall give special attention to philosophical views of apes.

Distancing and Distorting Animals

The answer to the question I have just formulated seems to be that we conceal to ourselves our cultural practices concerning animals in a number of ways. One of these 'distancing devices' is detachment: citizens nowadays have little contact with animals; caring for them is usually confined to certain species, and such common cultural activities as slaughtering and hunting quickly become looked upon with indifference when people grow up. Another strategy is active concealment. In Western societies nowadays, industrial breeding, killing and slaughtering tend to be hidden from public view. In addition, keeping very large numbers of animals in identical cages makes the individual animal anonymous, and our food terminology – 'beef', 'veal', 'hamburger' – tends to dissociate our food from the live animals as we know them. A third distancing device, most important in the present context, closely connected with the others, and deeply ingrained in our culture, is misrepresentation or ideological distortion. Representing animals as dull, brutish or even evil beings with no real subjectivity and feelings makes it easier to exploit them in a variety of ways.

A strict opposition between human and animal, implying a categorical boundary between both, is a widespread phenomenon in Western culture. It played a fundamental role in the articulation of middle-class cultural identity, which defines a civilised being as one who controls her or his animal impulses, and tends to look upon other social categories such as peasants or the working classes as closer to animality because they are, so the middle classes believed, less controlled in this respect. One culture-specific background to the way the human–animal boundary has usually been constructed in Western discourse, legitimising daily practices, is the Christian idea that human beings stand high above the animals because they are the only creatures created in the image of God. But there is another, more general reason for negative views of animals, which in fact the Western cultural tradition shares with many others: animals by their very nature behave in uncivilised ways – ways disapproved of in most human cultures. Therefore they are forceful symbols of uncivilised behaviour. Animals and animality are good vehicles for symbolising, for thinking, for moralising, for disapproval: 'you behaved like an animal'; 'the prisoners were treated like animals'. Of course there are differences between our ways of dealing with particular species. Usually, for instance, we refrain from eating animals who are close to us, such as dogs and other companion animals, while some other animals are killed for pleasure. Apart from carrying the usual animal connotations, many specific animals – rats, eagles, pigs, lions, etc. – also have specific symbolic meanings.

The Case of Apes

Apes, who for a non-indigenous species are quite prominent in Western cultural symbolism, form a special case. No animal has been so deeply involved in Western ideas on human nature, morals and origins. During the first years when Charles Darwin was pondering evolution and its mechanisms, he gave expression to his bewilderment with the following exclamation, jotted down in one of his notebooks: 'Our grandfather is the devil under form of baboon!' We can understand what he meant only if we realise that most of the various cultural roles apes and monkeys have played in European history have been negative ones. In medieval imagery, they were associated with sin and the devil, with frivolity, folly and hideousness, with impulsivity and wantonness. In modern times such connotations persisted. Against this background, Darwin's perplexity is easier to understand.

Until recently, apes, especially gorillas, were pictured as bloodthirsty monsters, and used to symbolise political opponents such as Bolshevists in the eyes of German fascists, or the threat of working-class communists in the eyes of the bourgeoisie. King Kong is but one recent avatar of a long tradition of Beauty-and-the-Beast stories. During the late nineteenth century, criminals and prostitutes were seen as suffering from an atavistic pathological backsliding into the brutish nature of prehistoric apemen. Monstrous apishness was also a way in which human motivation was conceptualised, not only in psychoanalysis, but quite generally: behaviour was seen as the outcome of a clash between our animal impulses, dating from primeval times when we were still wild beasts, and civilised control which had to tame the beast within. Apes have mostly been used to think and symbolise *with*, rather than being thought themselves in their own right, the way they are.

A short look at the roles of apes and monkeys in non-Western cultures may, like the imaginary world with which we began, serve to open our eyes to the peculiarity of the way in which we deal with nonhuman primates in the West. In medieval Japan, the Japanese macaque served as a revered mediator deity between men and gods – although in more recent times, it assumed the role of mocked scapegoat, standing for undesirable qualities, and that of a derided trickster/clown, challenging basic assumptions of Japanese culture. In traditional Chinese culture, gibbons were singled out as the aristocrats among apes and monkeys, symbolising the unworldly ideals of the poet and the philosopher, mediating between man and mysterious nature by initiating him into science and magic. A popular Hindu deity is Hanuman, the monkey-god, a central figure in the great Hindu epic Ramayana, worshipped in numerous temples, especially in southern India. In many tribal cultures, apes, monkeys and other animals are considered as persons and treated accordingly, even though they may be hunted and killed (as may human beings in some tribal cultures). When this happens gifts must be given and rituals followed.

Similarity and Ambiguity

Having seen that apes, compared with other animals, constitute a special case, we might ask why this is so. Why do apes carry such a heavy load of (usually negative) meanings, and play such a prominent role in human cultural imagery? I think this has to do with their great similarity to ourselves: apes are ambiguous beings because they resemble humans. They challenge the possibility of drawing a neat boundary line between humans and animals. They are neither completely human, nor completely animal, but both at once, or somewhere in between. They inhabit the margins of humanity. Now, as that branch of anthropology which studies the ways peoples in different cultures conceptualise and categorise themselves and their environments has demonstrated, the most suitable symbols are often drawn from those entities and situations which are ambiguous with respect to the classificatory grids we apply to reality.

Some examples may make somewhat clearer what this means. Things which are marginal or fall in the shadowy boundary area between two fields or classes of phenomena, such as dirt, menstrual blood or faeces, or animals living both on the land and in the water, often are very important in various cultural contexts. Their transgression of boundaries we think should be respected and the difficulty of pigeonholing and comprehending them makes such things uncanny and frightening. From among all that is available in the environment, such ambiguous entities are most suitable for expressing human meanings and feelings, and in many cultures it is precisely such entities which are chosen to be manipulated in rituals, to be subjected to taboos, or to curse and to insult people with. The uncanny, ambiguous ape, betwixt and between humans and animals, I hold, is suited as few other things are to express and signify what it is to be human.

The apes' ambiguous similarity to ourselves, including a whole repertoire of emotions, gestures and other behaviours we immediately recognise, makes them a potential threat to our own identity, and results in our complex reactions to this close relative of ours. This threat makes it necessary for humans to reaffirm vigorously the ape's brutish animality and low status, in order to protect the clearcut boundary between humans and animals. For this boundary is one that we need desperately, in order to be able to go on killing and eating millions of animals every year, while we refrain from killing or eating humans. By distancing devices the ape's disquieting familiarity is kept at a safe distance, spirited away, conjured into thin air.

Their ambiguous similarity to ourselves, their mirror-quality, makes the apes especially suited to play the role they are intended to play in the Great Ape Project: that of a bridgehead into the realm of nonhuman animals, which helps us to overcome the traditional wide gap between humans and animals, and to extend, beyond the biological boundaries of our species, the moral community of which we consider ourselves part.

Ignoble and Noble Apes

So apes are good to think and to symbolise with. They lend themselves to the expression of a great variety of meanings, often, as we have seen, negative ones: the immoral, unacceptable aspects of our own behaviour. But this was not always so. Eighteenth-century pictures of apes, for instance, show peaceful, rather human-looking creatures living a happy, natural life, quite unlike the ferocious, bloodthirsty monsters we encounter in the second half of the nineteenth century and most of the present century. This new image of apes probably has something to do with a changing view of nature, which during the last century came to be seen as the product of a hard-fought struggle for life in a quite literal sense, instead of a harmonious Great Chain of Being created by the Almighty. In addition, in the context of colonial expansion, apes, and especially the gorilla, came to be seen as powerful personifications of wildernesses to be fought and conquered heroically by civilised man. The standard procedure for obtaining apes for museums and scientific research was to go out and shoot them. Apes destined to be kept alive in zoos, as companion animals and as laboratory animals were – and indeed still are – preferably captured when still young, by shooting the mother.

A significant change in the way people in the West felt about apes took place during the 1960s, when the first field studies of free-living apes began. A typical, and for our present sensibilities most shocking, example of how things were until then is pictured in a scene from a colonial propaganda film made in the Belgian Congo on behalf of the Belgian government. The film, which circulated widely in Belgian cinemas during the 1950s, shows in great detail how Belgian scientists shoot and kill an adult female gorilla who is carrying a gorilla baby. The adult female's body is then skinned and washed in a nearby stream, while the crying gorilla child, destined for a Belgian zoo, is sitting next to it. Ten or fifteen years later such a scene, in a film meant to be seen by Western families with their children, had become unthinkable.

The new attitude was one of sympathy with, and admiration for, the harmonious, natural life of the chimpanzees studied by Jane Goodall and others, brought to the public by *National Geographic* and by television. The publication, in the late 1970s, of new data on brutal violence, killing and cannibalism among chimpanzees therefore came as a shock to public opinion. Chimpanzees turned out to be less gentle, loyal, loving and noble after all. The cultural role which came to be played by apes during the 1960s is similar to, for instance, that played by little birds in nineteenth-century children's books, forming pairs, building cosy nests and carefully rearing their children together – the ideal of the bourgeois family. Every period has its favourite animals, its ideals of living in harmony, its ethical models for human action.

The Great Ape Project is symptomatic of a new, positive attitude towards animals. This new sensibility has certainly been influenced by field studies on ape behaviour, but there is more to it. A more general background is to be found in profound changes in Western views of human beings and the world. The traditional idea of our unique dignity as the sole creature created by God in his own image has lost much ground, not least to the secular Enlightenment view of humans as uniquely rational and Western culture as the natural goal and outcome of world-historic progress. The human–animal boundary was then threatened in the nineteenth century by the rise of the theory of evolution and the idea of a physical instead of a metaphysical origin of humankind; but that threat was quickly warded off by turning a descent from apes into an ascent towards civilisation. Now the critique of many aspects of Enlightenment ideas has left us without a clearcut, unequivocal definition of our place in nature and without a concomitant legitimation of our behaviour, including that towards other animals.

What seems to be happening is a constant extension, in concentric movements, of the group of creatures we mean when we say 'we', creatures we consider as fellows and as moral equals. One step has been the abolition of black slavery, another one the end of colonialism, still another one the emancipation of women. The extension of the community of our moral equals beyond the biological limits of our species fits into this pattern.

The Philosopher's Ape

More often than not, animals in general, and nonhuman primates in particular, have been stereotyped as low, brutish beings, and excluded from the community of beings worthy of moral respect in the same degree as humans. An important historical background to this attitude was the conviction that there is an absolute, rather than relative, distinction between humans and animals, one of an essential nature, not of gradual differences, to be found in the human mind. In the foregoing, we have been assuming that this is an incorrect view, and that therefore animals deserve more respect as knowing and feeling subjects. But is that really so? The presupposition that there are only gradual differences between humans and animals may seem plausible at first sight, perhaps even self-evident, but it is in fact contested and controversial, especially among philosophers.

While in the English-speaking world most philosophers would subscribe – or would at least be inclined to do so – to the idea of perhaps large, but ultimately gradual differences between humans and animals, most philosophers from the European continent would not underwrite the continuity of beasts and humans. A considerable number of continental philosophers operate in the wake of Aristotle or Descartes rather than that of Locke or Hume. They engage in Kantian criticism rather than evolutionary epistemology, in phenomenology or hermeneutics rather than naturalistic philosophy of mind. As different as these continental philosophical outlooks may be, they do have one thing in common: they all, in one way or another, draw a strict boundary line between animals and humans, and assume the gap between both is unbridgeably wide.

They do so because, in the process of analysing according to their specific methods, they encounter a characteristic which in their eyes is uniquely human – reason, mind, rationality, intentionality, self-consciousness, or whatever term they use for it. They all see no possibility of fully accounting for this characteristic in terms of gradual differences or continuity with characteristics found in animals, the central nervous system, organic processes in general, or indeed anything three-dimensional and physical. In their eyes, the human, rational, self-conscious mind is a qualitatively different, irreducible phenomenon that gives to the entity which possesses it a very special place in nature compared to those who do not.

One typical and influential advocate of this point of view was the phenomenologist Max Scheler, who during the 1920s – like his German colleagues Martin Heidegger and Helmuth Plessner – tried to make sense of the first experiments on the cognitive abilities of chimpanzees, conducted by the German biologist Wolfgang Köhler. Scheler argued that although chimpanzees are intelligent, their behaviour and perceptions are still determined directly and fully by their instinctive impulses and needs. Therefore, they are not 'open to the world' – *weltoffen*. They have not entered that dimension of existence in which it is possible to know the (things of the) world as such – cut loose from meanings lent by instincts, such as being edible, being dangerous or providing shelter – and, concurrently, in the same movement of mind, to know oneself as such. Conscious apes, Scheler holds, are not present to themselves in the way self-conscious humans are, and their behaviour comes about in a mechanical way rather than by free choice.

Philosophical positions of this type are defended by many, with sophisticated and elaborate arguments. It is too easy simply to discard such interpretations of animal subjectivity as an ideology justifying our exploitation of animals. Addressing and refuting the arguments is a better strategy. Scheler's argument that apes do not know the things of the world as such, for instance, is, among other things, based on the assumption that they are not able to transfer or translate information from one sensory modality (e.g. auditory) to another (e.g. tactile or visual), or to integrate information available in several sensory modalities into an awareness of one underlying thing in the world to which all these different data pertain. Recent research, however, has shown that apes are very well able to make such transfers, which refutes at least this line of argument. As to its moral side, even if philosophical positions such as Max Scheler's turn out to be right, it does not follow automatically that humans deserve more respect because they are rational and self-conscious.

Most of traditional philosophy has not reflected upon real animals, but upon malignant stereotypes which turned the animals into brutish monsters. Therefore it is necessary to rethink our conceptions of animals, especially apes, philosophically and morally, in the light of new empirical knowledge now available showing that apes are more similar to us than we have ever realised and would ever have dared to realise. Apes may well turn out to be human in terms of several traditional components of that concept, such as the possession of self-consciousness and a free will.

The French philosopher Emmanuel Levinas, reflecting upon the way we experience the other person's gaze, has criticised traditional Western philosophy for taking the subject, the I, as the absolute point of reference for all other beings. Could this view of Levinas be, we might ask, another case of anthropocentrism, and ironically so, because, while the intention is to think the irreducibility of 'other' to self, our animal 'others' are disregarded once again? Is the gaze, which, according to

Levinas, appeals directly, without any mediation, to our moral awareness, different when it is not that of a human child but that of a gorilla
child or an orang-utan child?

Perplexities

Reflections upon how we have treated, and still do treat, other animals
leads to feelings of uneasiness, if not perplexity. For one thing, how far
should we go in extending the boundary of the community of moral
equals among which we count ourselves? Is the fact that another being
has an emotional life and a subjectivity similar to our own a good reason
for respect? Or does respecting what is similar to ourselves rather than
different from ourselves imply that we once again take our own, human
nature and our allegedly unique dignity as the pivotal point of reference,
as an absolute standard against which to measure everything else? In
that case we would only be defending a new form of anthropocentrism.
How, for instance, should we behave towards squid living in the dark
depths of the Pacific Ocean, towards lobsters, spiders, rats?

Why, to address another perplexing question, should the view of the
interior of a slaughterhouse in a land of intelligent, highly civilised,
gentle pigs, with the carcasses of hundreds of slaughtered humans, split
into two and hung up, be *more* shocking than similar views of slaughtered pigs which are readily available near every large city in the
Western world? The reasons may well be psychological and sociological
rather than moral ones, and have to do with our concern for our own
individual and communal well-being and with the necessity to maintain
social order. That we say we may not kill or eat each other may have to
do with our fear of being killed or eaten ourselves. Admitting the moral
equality of other animals like apes, pigs or cows seems to leave us with
only two possibilities: being prepared to rear, kill and eat humans as
well, or refraining from killing and eating equals altogether. What we
eat and what we do not eat, what we kill and what we do not kill, what
we define as cannibalism or murder and what falls outside this definition – here our most fundamental taboos are involved.

Do we really know ourselves? Traditional conceptual frameworks –
such as the Christian world view, or Enlightenment ideas on progress,
rationality and civilisation – which defined our place in nature, legiti-
mised our behaviour and warded off threats to the animal/human
boundary, have lost most of their power. How can we act morally after
the implosion of our traditional narratives? Perhaps our nearest relatives in nature, the apes, who problematise, destabilise, and thereby
renew our identity, can help us, by playing the role of missing link
between humans and animals. We could begin by taking seriously that

fundamental, primordial experience we have when confronted with them: an experience of other subjects, other persons, other individuals, accompanied by the strong moral feeling that they deserve as much respect as we do ourselves; an experience which is hampered less and less, these days, by mechanisms that keep animals at arm's length and distort the way we spontaneously experience them. A movement has started, and this book is part of it, from thinking with apes – employing them, that is, as vehicles of meaning in human discourse – towards thinking about them as, and respecting them for, what they are themselves.

Further Reading

1. Mary Douglas, *Purity and Danger* (London, 1966).
2. Robert van Gulik, *The Gibbon in China: An Essay in Chinese Animal Lore* (Leiden, 1967).
3. Donna Haraway, *Primate Visions: Gender Race and Nature in the World of Modern Science* (London, 1989).
4. H. W. Janson, *Apes and Ape Lore in the Middle Ages and the Renaissance* (London, 1952).
5. Mary Midgley, *Beast and Man: The Roots of Human Nature* (London, 1980).
6. Emiko Ohnuki-Tierney, *The Monkey as Mirror: Symbolic Transformations in Japanese History and Ritual* (Princeton, 1987).
7. James Serpell, *In the Company of Animals* (Oxford and New York, 1986).

13

Spirits Dressed in Furs?

ADRIAAN KORTLANDT

The first time a free-living chimpanzee and a human scientific observer gazed at close range into each other's eyes, the human eyes were those of Adriaan Kortlandt. Born in Holland, Kortlandt studied – both before and during the war – the behaviour of cormorants in the wild for his doctorate at the University of Amsterdam. He also participated in the underground resistance during the Nazi occupation. Some time after the war he became professor of animal psychology and ethology. He pioneered the detailed observation of free chimpanzees in their natural environment in Africa. He has published extensively in scientific journals on the nature of instinct, on his fieldwork with chimpanzees, and on an ecosystem approach to problems of evolution in humans and apes. In this essay he considers our knowledge of the minds of the great apes in the light of his own experiences – and also in the light of our tendency to harbour prejudicial views of those who are not like us.

In olden times the world was haunted. Mountains, rivers, forests and seas were inhabited by gods, demons and all kinds of spirits which could be friendly or angry, and would cause earthquakes, thunderstorms or a good harvest, just as their moods dictated. Little by little, in the course of time, the picture changed. The world gradually became despirited. Aristotle still attributed a vegetative soul to plants, a sensitive soul to animals and a thinking soul to humans. The idea still survives to some extent in Roman Catholic philosophy. Reformation theology was more stringent and denied the presence of a spirit or soul in animals, though Luther is said to have written that dogs would be admitted to Heaven. (He had a dog, of course, and would not have liked to leave him behind.) In 1649 Descartes made the next step and declared animals to be machines, not only without an immortal soul but even without any mental experiences.[1] More than two hundred and fifty years before Pavlov[2] he described stimulus conditioning of dogs and stated that 'les bestes n'ayent point de raison, ny peut estre aussi aucune pensée', but nevertheless 'on peut avec un peu d'industrie changer les mouvemens du cerveau dans les animaux depourveus de raison'. (Beasts have neither reason nor perhaps any thought [but] with some ingenuity one can change the brain movements in them.)[3] His student Malebranche drew the logical conclusion and flogged his dog savagely. When his neighbours objected he told them that according to Descartes animals are inanimate machines, and asked: do you protest if I beat a drum?

In the follow epoch, Cartesian philosophy set the tone in scientific as well as theological thinking on animal minds, usually called souls: science and religion had not yet divorced.[4] Eventually, in our century, Pavlovian and Watsonian behaviourists pushed the idea to its limit and developed a human psychology without a mind. Similarly, Lorenzian ethologists declared animal minds, if existing, to be unknowable.[5] To complete the picture, some theologians proclaimed God to be dead. All the ghosts were leaving. Academic thinking was becoming mindless, on Earth and in Heaven. The legal implication would be that henceforth no one could be liable to conviction for cruelty because suffering could not be proven scientifically, in humans and in animals.

There were other trends. In 1760 a German professor, Hermann Samuel Reimarus, published a bestseller: *Allgemeine Betrachtungen über die Triebe der Thiere* (General Considerations on the Instincts of Animals).[6] In this book he rejected Descartes' concept of animal-machine and argued (translated into modern terminology): (1) animals have sense organs similar to ours which conduct stimuli to the brain; (2) they organise their behaviour according to their perception of the world around them; (3) this proves that they have a mental representation of the world around them; (4) that is, they have a '*Seele*' (in German, the psychological concept of mind and the theological concept of soul are homonymous). He also recognised, among other things, that quadrupedal animals have dreams. Thus he was 100 years ahead of his time on the homology argument of Darwinian animal psychology, and 160 years ahead of von Uexküll's *Umweltlehre*[7] (see below).

All Hell broke loose with this blasphemous postulating of an animal '*Seele*'. However, Cartesianism retained the upper hand until 1859, when Darwin's *Origin of Species* appeared. Then the recognition of evolutionary homology, including humans, led to the tripartite distinction: comparative anatomy studied the structures; comparative physiology the functions; and comparative psychology the minds of animals, including humans. This was the heyday of subjectivistic interpretation of animal behaviour, and it was carried out *ad absurdum*.[8] The Deutsche Gesellschaft für Tierpsychologie officially endorsed the calculating horses and thinking dogs, including a dog who at Christmas time told his mistress by paw-spelling that he believed the infant Jesus was coming.[9] After the unmasking of the affair, the embarrassment of such eminent believers as von Buttel-Reepen, Edinger, Haeckel, Plate, Sarasin and Ziegler made the term 'Tierpsychologie' unspeakable in Germany. Ethology was created instead and eventually was exported all over the world.

Another development was equally undermining for subjectivist animal psychology. The American philosopher William James (1890) had defined psychology as 'the science of mental life, i.e. 'the study of the stream of thought', or 'of consciousness'.[10] Freud, however, produced evidence indicating that many, or even most, of our psychological processes proceeded unconsciously.[11] To find out which processes worked consciously and which unconsciously one had to interview the subject and, by means of ingenious procedures, analyse the hidden meanings behind his or her dreams, blunders, symbols, etc. The unconscious had become accessible, that is, in humans. Animals, however, might very well behave unconsciously too.

Psychoanalytical psychology inspired, to some extent, work on animal behaviour (e.g. in the discovery of displacement activities). Interviewing animals, however, remained impossible. At that time, apes' ability to use sign language had not yet been discovered. Ethologists continued to assert that the mental life of animals was unknowable. Scientifically the word 'unknowable' is, of course, disastrous because it blocks research into the unknown. No one can predict, for instance, whether it will *forever* remain impossible to link an animal nervous system to a human one.

Should we *for the time being* continue to believe that the inner experiences of animals remain inaccessible to us, except perhaps by means of ape sign language? The consequences would be considerable. A conviction for cruelty to animals based on expert testimony would be legally impossible, except where apes were involved. The general public, and dog and cat owners particularly, would not accept such expertise. Ethologists themselves would be extremely unhappy too. Small wonder that Griffin made a desperate attempt to revive scientifically the animal mind.[12] His endeavour was, however, naive and epistemologically untenable. For instance, Freud was not mentioned, nor were such phenomena as subliminal perception.

Besides objectivistic and subjectivistic interpretation of animal behaviour, there is a third way. When we humans are happy, or sad, or in love, or angry, we are not just happy, sad, in love or angry *per se*: we are happy or sad *about* something, and we are in love or angry *with* someone. When we see colours we do not simply see them as images on our retina but as colours of flowers, leaves and other things *around us*. When we think, we do not think *in vacuo* but we think *of* things or people. That is, our mind is not located in our brain but in the world around us. The nervous system is only a tool for creating a picture of the world. This subjective world *is* our mind. Similarly, animal behaviour should not be seen as an outcrop of inner experiences or driving forces inside the animal, but as the fabric of meanings which are projected into the outer world by the animal. This includes what von Frisch incorrectly called sense physiology[13] (it is not physiology), what von Uexküll called *Umweltforschung* or *Umweltlehre*[14] (which I tried to translate as 'animal cosmology') and a part of what is today called cognitive psychology. It is a kind of science that is rather alien to classically trained comparative psychologists and ethologists. Neither von Frisch nor von Uexküll, nor I myself, met with much response among behaviourists and ethologists when we presented these ideas, and cognitive psychologists working on animals do not, as a rule, publish their findings in ethological journals. Yet I think that research on the subjective worlds of animals holds the epistemological clue to the most fundamental problem of understanding animals.[15]

At this point we may consider the problem of human perception of differently disposed human beings. In medieval times, the mentally disturbed were chained and kept in cages or prison-like madhouses. They were thought to be possessed by demons. On holidays they were displayed to the public for entertainment and to be teased. After the French Revolution of 1789 had proclaimed *liberté, égalité et fraternité*, the Parisian asylum director Pinel started in 1792 a programme of unchaining his patients, treating them humanely, giving them (restricted) liberty and applying verbal therapies.[16] Humanisation of the 'subhumans' had begun. Yet these ideas permeated very slowly into society. When I was young, elderly people told me that in their youth it was quite common to deride and to tease mentally and physically handicapped people openly in the streets and in public places. The popularisation of psychiatry by the media started only in the 1920s, after Freud's writings became common reading among the educated.

Even then, class society and colonialism maintained ideas about genetic, racial and cultural superiority/inferiority. In 1928, when I was a schoolboy, 'Negroes' and Amerindians were still dressed up and trained to behave as savages to be shown at an official exhibition of industry in Holland. General awareness of what was wrong came to Europe and the United States only after six million so-called *Untermenschen* (subhumans) had been killed in concentration camps, and after colonialism had broken down. Publicly expressing racism became forbidden by law in some Western countries. However, teasing the apes, monkeys and large carnivores in zoos remained commonplace. It was the last outlet for instinctual inter(sub)specific anti-competitor and anti-predator behaviour among humans; a similar instinct to that causing dogs to chase cats.

It may be significant that soon afterwards general attitudes towards primates began to change. Large-scale primate behaviour research in zoos and in the wild started around 1959, i.e. after the Darwin centennial commemoration at the International Congress of Zoology in London in 1958. (Köhler and Yerkes had remained lonely voices in the wilderness.)[17] Later, in the 1960s, a young colleague told me that he had seen two leading elderly colleagues in the behavioural and evolutionary sciences in front of the chimpanzee cages at London Zoo: they were acting just like the visitors at a madhouse before the French Revolution. He felt deeply shocked and ashamed. I must admit that on hearing his story, I was also shocked because I realised that I was halfway between the grandfather and grandson scientific generations, and could empathise with both. How could I? I had been lucky enough in May 1960 to be the first human being to observe chimpanzees at close quarters in the wild. These were not the dirty and neurotic creatures whom I knew from zoos and laboratories. These alert and animated, but insecure and unsettled, free-living creatures were interested in anything unusual or remarkable – from a beautiful sunset to a piece of barbed wire. They continuously varied their habits and behaviour. They always hesitated before deciding which direction to take or which fruit to eat. They came to my hide and their piercing brown eyes met my grey ones: scratching themselves with wonder, they then walked away without having solved the mystery. They seemed to have lost the certainty of instinct but not gained the certainty of knowledge. These were not animals, nor humans either, but eerie souls in animal furs. A cold shiver went down my spine. And then they occasionally signed to one another by means of a hand or arm gesture. Again a shiver went down my spine. It was the greatest experience of my professional life.

Established attitudes and habits die off very, very slowly. Until the late 1960s at least, apes and monkeys were still frequently branded by scientists as subhuman or infrahuman primates – *Untermenschen* – rather than as nonhuman primates. They were, and still are, often treated accordingly. In our present time, chimpanzees and orang-utans are still regularly abused as clowns in TV advertisements apparently without raising a public outcry. When Bert Haanstra's film *Ape and Super Ape* or Hugo van Lawick's film *People of the Forest* are shown, even an audience of professional primatologists will roar with laughter when chimpanzee behaviour is shown. Yet these people know very well that these free-living chimpanzees behave quite naturally and are adapted to their environment. There is nothing funny about this, except when the chimps themselves make fun. These human laughers still perceive chimpanzees as silly idiots and circus clowns dressed in animal furs, impersonating human follies, rather than as creatures in their own right.

Our scientific image of the great apes in general, and of chimpanzees in particular, has taken on a new dimension in recent years. In the past, chimpanzees were depicted as peaceful fruit-pickers hanging and swinging by their long arms in the forest canopy.[18] We now know that many of them inhabit open savannah landscapes (a fact scientifically known since 1930 but mostly ignored for about half a century), that they regularly walk bipedally in the wild, and that they hunt and consume fairly large prey. They also use tools for various purposes, and primitive weapons against natural enemies. Furthermore they deceive one another, rape females, practise power politics, occasionally kill and cannibalise, make war against alien groups and (in one case observed in captivity) castrate the bullying boss.[19] Similar data are beginning to emerge on the gorilla and the orang-utan. It is a very grim picture. As grim as prehistory, cultural anthropology and world history. Our ape science is no longer beyond good and evil.

I, for one, have always believed that the real chasm between 'animals' and 'humans' was between, on the one hand, the great apes and, on the other hand, the lesser apes, baboons and monkeys. Recent research has suggested that we should perhaps locate the gulf below the baboons.[20] To be conservative, for the time being it seems proper to accord at least the great apes the status of less-gifted humans and to treat them accordingly.

I am aware, of course, that living and sometimes non-anaesthetised subjects are needed in certain medical experimentation aiming to alleviate the suffering of humans. Those who have seen what is going on inside a hospital, and those who have lost a loved one owing to the impotence of medical science, will understand what I mean. I myself have seen some heart-breaking research in primate centres, particularly in the psychological and psychiatric field. However, as a student of psychology I have also seen enough in mental wards to appreciate the value of such research. On the other hand, how can we justify such research with our innocent ape cousins, while doing so is not allowed even with those humans who are guilty of the most horrifying crimes against humanity?

Notes

1. R. Descartes, *Les passions de l'âme* (1649; available in several English editions).
2. I.P. Pavlov, *Twenty Years of Experiments in the Objective Study of Higher Nervous Activity (Behaviour) of Animals* (in Russian) (1923; English edition: *Conditioned Reflexes* (Oxford University Press, London, 1927)).
3. Descartes, *Les passions de l'âme*.

4. For example, A. le Grand, *Dissertatio de carentia sensus et cognitionis in brutis, Londini* (1675; English edition: *An Entire Body of Philosophy, According to the Principles of the Famous Renate Des Cartes, 3, A Dissertation of the Want of Sense and Knowledge in Brute Animals, Giving a Mechanical Account of their Operations* (Blome, London, 1694; and Johnson Reprint Corp., 1972)).

5. For example, N. Tinbergen, *The Study of Instinct* (Clarendon Press, Oxford, 1951).

6. H.S. Reimarus, *Allgemeine Betrachtungen über die Triebe der Thiere, hauptsächlich über ihre Kunsttriebe* (Bohn, Hamburg, 1760; French edition published in 1770).

7. J. von Uesküll, *Theoretische Biologie* (Paetel, Berlin, 1920; revised edition: Springer, Berlin, 1928; imperfect English edition: *Theoretical Biology* (Paul, London, 1926)).

8. See, for example, G.J. Romanes, *Mental Evolution in Animals* (Kegan Paul, London, 1883).

9. K. Gruber, 'Vom denkenden Hunde Rolf', in H.E. Ziegler (ed.), *Die Seele des Tieres* (Junk, Berlin, 1916), pp. 87–99: H.E. Ziegler, 'Mitarbeiter der Gesellschaft für Tierpsychologie', *Mitteilungen der Gesellschaft für Tierpsychologie*, vol. 1 (1913) pp. 3–4.

10. W. James, *The Principles of Psychology* (Holt, New York; Macmillan, London, 1890).

11. S. Freud, *Die Traumdeutung* (Deuticke, Vienna, 1900; available in several English editions); S. Freud, *Zur Psychopathologie des Alltagslebens* (Karger, Berlin, 1901; available in several English editions).

12. D.R. Griffin, *The Question of Animal Awareness* (Rockefeller University Press, New York, 1976).

13. K. von Frisch, 'Der Farbensinn und Formensinn der Bienen', *Zoologische Jahrbücher, Abteilung für allgemeine Zoologie und Physiologie der Tiere*, vol. 35 (1915) pp. 1–182.

14. Uexküll, *Theoretische Biologie*.

15. H. Hediger, *Wildtiere in Gefangenschaft* (Schwabe, Basel, 1942; English edition: *Wild Animals in Captivity* (Butterworth, London, 1950)): I. Kant, *Critik der reinen Vernunft* (Hartknoch, Riga, 1781; available in several English editions); A. Kortlandt, 'Cosmologie der tieren', *Vakblad voor Biologen*, vol. 34 (1954) pp. 1–14 (English translation in preparation); K. Lorenz, *Die Rückseite des Spiegels* (Piper, Munich, 1973; English edition: *Behind the Mirror* (Methuen, London, 1977)); Uexküll, *Theoretische Biologie*.

16. P. Pinel (1792), in S. Pinel, 'Bicêtre en 1792. De l'abolition des chaînes', *Mémoires de l'Académie Royale de Médicine*, vol. 5, no. 2 (1836) pp. 31–40.

17. W. Köhler, *Intellegenzprüfungen an Anthropoiden, I. (Einzelausgabe)* (Königliche Akademie der Wissenschaften, Berlin, 1917; English edition:

The Mentality of Apes (Harcourt, New York, 1925)) (note that the English translation understates his point on mental representations (*Vorstellungen*)); R.M. Yerkes, *Chimpanzees* (Yale University Press, New Haven, CT, 1943).

18. For example, W.E. Le Gros Clark, *History of the Primates*, 1st to 10th editions (British Museum (Natural History), London, 1949–70).

19. See, for example, A. Kortlandt, 'Marginal habitats of chimpanzees', *Journal of Human Evolution*, vol. 12 (1983) pp. 231–78.

20. R. Byrne and A. Whiten (eds), *Machiavellian Intelligence* (Clarendon Press, Oxford, 1988).

IV

Ethics

14

Apes, Humans, Aliens, Vampires and Robots

COLIN MCGINN

Colin McGinn taught philosophy at University College, London, before becoming Wilde Reader in Mental Philosophy at Corpus Christi College, Oxford. Since 1990 he has been distinguished professor of philosophy at Rutgers University, New Jersey. His research has concentrated on the nature of mind and consciousness, and his books include The Character of Mind, Wittgenstein on Meaning, The Subjective View *and, most recently,* The Problem of Consciousness. *He also writes fiction: the contribution that follows draws philosophical conclusions from a fictional scenario.*

As a child, you tend to take your position in life for granted, as written into the natural order of things. You were born, say, into a white middle-class family, you are comfortably off, in good health and not in any particular distress. You have rights and privileges, and these are generally respected. You aren't hungry or imprisoned or enslaved. You go on nice holidays. At an early age, you probably assume everyone lives like this. It seems natural that you enjoy the kind of life providence has granted you. You don't think about it.

Then you start to notice that others are less fortunate (and some others more fortunate). You see people around you who are poorer than you, possibly homeless, or who have something serious mentally or physically wrong with them. You start hearing about people in foreign countries who are starving to death, or being blown up in wars, or suffering from terrible diseases. Some of them are children like you! These facts jar on you; and they force you to make comparisons with your own life. Soon you are struck with a certain terrifying thought: that it is really just *luck* that you are not in their shoes. You happen to have been born into a certain class, in a certain part of the world, with certain social arrangements, at a certain period in history. But there is nothing necessary about this – it is just the luck of the draw. Things could have been different in ways that don't bear thinking about. You ask yourself what your life would have been like if you had drawn the short straw and lived in less felicitous circumstances. You imagine yourself born into a land of famine, or arriving on the scene before medicine made any progress with plagues, or before modern plumbing. You thus entertain a kind of philosophical thought: that it is just *contingent* that things are as they are, and that you could have been very much worse off. You are just lucky. Equivalently, you see that it is just bad luck for the others that their lives are as hard as they are. There is no divine necessity or inner logic about any of this. It is basically a moral accident. There but for fortune . . .

And with this thought social conscience begins. Since there is no deep necessity about the ordering of well-being among people, we should try to rectify (avoidable) inequalities and misfortunes. The arbitrariness should be removed from the distribution of well-being. We should discover the sources of misery and deprivation and try, where possible, to erase them. We should certainly not voluntarily contribute to the disadvantaged position of others. We should not exploit the power that is ours by sheer cosmic luck. Thus, morality is founded in a sense of the contingency of the world, and it is powered by the ability to envisage alternatives. Imagination is central to its operations. The morally complacent person is the person who cannot conceive how things could have been different; he or she fails to appreciate the role of luck – itself a concept that relies on imagining alternatives. There is no point in seeking change if this is the way things *have* to be. Morality is thus based on modality: that is, on a mastery of the concepts of necessity and possibility. To be able to think morally is to be able to think modally. Specifically, it depends upon seeing *other* possibilities – not taking the actual as the necessary.

I think, to come to the present point, that human adults persistently underestimate the role of biological luck in assuring our dominion over the rest of nature. We are still like children who take the contingent facts to be necessary, and thus fail to understand the moral significance of what actually goes on. People really do believe, in their bones, that there is a divine necessity underwriting our power over other species, so they fail to question this exercise of power. Indeed, this assumption is explicitly written into many religions. In every possible world we are at the top of the biological tree. As children, we naively took our family position to be the locus of cosmic necessity; now we assume that our species position is cosmically guaranteed. We assume, that is, that our relation to other species is basically the way things *had* to be, so that there is no point in questioning the ethics of that relation. Hence social conscience stops at the boundary of the human species, give or take a bit of supererogation here and there. We don't take seriously the idea that it is just luck that our species is number one in the biological power hierarchy. So our conscience about our conduct in the biological world isn't pricked by the reflection that *we* might have been lower down in the scale of species domination. We therefore need to bring our species morals into line with the real facts of biological possibility.

To be specific, we fail to appreciate that we could have been in the kind of position with respect to another species that apes now occupy with respect to us; so we protect ourselves from the moral issues that arise about our actual relation to apes. Or rather, we acknowledge the contingency of our biological position in odd and localised ways – as if our unconscious recognises it only too well but we repress it in the interests of evading its moral consequences. For our instinctive species-ism wavers when we consider ourselves on the receiving end of another species' domination. We allow ourselves to enter into this contingency in certain special sealed-off imaginative contexts – not in the world of hard moral and political reality. Significantly, these contexts typically involve horror and fear and loss of control. For the most part, now-adays, they take place in the cinema. I am thinking, of course, of science fiction and horror films. Here alternatives to our biological supremacy are imaginatively explored. Let me mention three types of fiction in which we humans assume a position of species subjugation – or contrive to escape such a position against considerable odds.

First, of course, there are the invading aliens from outer space, who come to destroy or parasitise or enslave the human species – the body-snatchers, stomach-busters and mind-controllers. Here the thought is that only space protects our species from the depredations of more powerful beings, so that space travel is a potential route to species demotion. Sheer distance is the saving contingency here. It is just luck that those aliens don't live on the moon, or else we would be their playthings even now.

Then there are the vampire stories, in which the theme of using the human species for food is paramount. A colony of vampires lives off the human inhabitants of a certain area, drinking their blood, killing other humans who get in their way. The humans are just a herd for the vampires. Usually the vampires are depicted as extraordinarily evil, gloating over the soon-to-be-punctured necks of their beautiful young victims, but sometimes they are portrayed more sympathetically, as just doing what nature designed them to do – slaves to their own biology, as it were. In any case, they are perceived as a terrifying threat to humans, and there is generally a good deal of luck involved in fending them off. It was a close thing that the entire human species wasn't condemned to be vampire-feed for all eternity. And it is lucky that we, the viewers, weren't born in Transylvania.

A third category of human demotion introduces machines, our machines. I suppose Frankenstein's monster comes into this category, since it was constructed by a human, albeit from organic parts; but a more recent example of the genre is the *Terminator* movies, in which the international computer network controlling nuclear weapons achieves self-consciousness one day and, fearful for its own survival at human hands, begins to wage war on its human creators, with very nasty consequences. This computer constructs its own formidable robots ('terminators') whose mission is simply to kill as many humans as possible, and they will not stop. This, then, is a case in which our artifacts rise up and exert domination over us, bringing untold havoc and misery to our species. And here the contingency is merely the level of technological advancement of our machines. If we are not careful, the message goes, our technology will come back to oppress us; so we had better not rely on luck to prevent this happening in the future. In fact, if time travel is possible, we should be thinking about it now, since the future may contain the very terminating machines made possible by extensions of our present technology. So, at least, the movies suggest.

Well, this is all good entertaining fun, but the point I want to make is that these nightmare fantasies represent, in sublimated form, our repressed sense of the contingency of our biological supremacy as a species. They are saying, '*You* could be in the position that other species are actually in – that you *put* them in.' And, of course, we are supposed to sympathise with ourselves in these possible fantasy worlds: we applaud the freedom fighters who seek to liberate us from the selfish domination of other kinds of being. We certainly don't think that might is right in *these* battles between the species. We have to fight them precisely because they are morally blind to what they are doing to us, or just outright callous. What I am suggesting now is that we take seriously the notion that we might have been, or could be, in such a position, and ask ourselves what moral principles we would want to see observed if indeed we were the weaker species. That is, we need a species morality informed by the idea of biological luck. Equivalently, we need to ask ourselves what rights need to be granted to species who *happen* to be thus subservient to us – apes in the present case. How does it look from their point of view? If humans had never evolved, then there would have been no scientific experimentation using apes as subjects, no confinement of apes in zoos and elsewhere, no systematic killing of apes for sport. Apes would undoubtedly have been better off without us. They are cosmically unlucky in the way *we* would be if any of the above nightmares become reality. And just as we would fight to have the evil effects of such bad luck reversed in our case – using sound moral argument as our justification – so we should recognise that the bad luck of apes in having humans to contend with should not be allowed to continue unchecked. In short, we should stop oppressing them. We should accord them the rights their intrinsic nature demands, not just acquiesce in the abuses of power consequent upon our chance biological supremacy. We might have been the ones in the cages or on the vivisection tables: and it is a cast-iron certainty we would not have liked it one bit. Morality, in short, should not be dictated by luck.

Let me end with an idea for a screenplay. We are a couple of million years into the future, and time has not been kind to the human species. Human intelligence reached a plateau in the twenty-first century, when the physiological constraints of giving birth stopped infant heads getting any bigger. Unluckily, too, the diseases of the modern world – physical and psychological – were not vanquished, leaving humans a generally sickly and neurotic lot. The pollution, the overeating, the crime, the stress have made humans a weak and enervated species. However, the apes have enjoyed a steady march forward. Their frontal lobes have been expanding all the time, they are fit and robust, and they have long since thrown off their human shackles. They have all the trappings of civilisation. Now, in fact, the status quo has been reversed: humans are now vulnerable to their whims. Some of the more unscrupulous of the gorillas – the ones with the flashy houses and private jets – have gone into business selling human specimens for a variety of purposes, no questions asked. Some go for medical experiments designed to benefit apes, others to slaughterhouses, the lucky few become pets, yet others are sold for interspecies prostitution. So far this is all illegal, done on the black market, and is officially frowned upon by the apes' government. But it is easy to arrange, given the vulnerable state of so many humans. The big problem, for the ape entrepreneurs, is getting the trade in humans accepted and legalised, so that they don't have to operate on the wrong side of the law. There is this annoying ape lobby, you see, that disapproves of subjugating humans in these ways, and, of course, the humans are less than thrilled about it themselves. The shady businessapes are working on the corruption of some high officials to get them to pass a law allowing what is now only done illegally. The propaganda, thankfully, is a breeze, given what all apes know about their treatment at the hands of humans for so many centuries – it is there in the history books. Serves them right, does it not? It looks like they are going to succeed in institutionalising their exploitation of humans, unless that brave coalition of good apes and desperate humans can prevent them . . .

OK, my point is this. Suppose this story became reality: wouldn't it be better to be able to say to the apes, who are generally a kind and decent species, that we stopped exploiting them voluntarily in the last decade of the twentieth century? We saw the error of our ways, so why should they repeat our earlier mistakes? We were not simply forced, by their biological ascendancy and our decline, to grant them rights in the middle of the 1000th century, say, after a bloody war; we just did it from moral principle well before we could be made to. We could thus appeal to their moral sense by citing our own earlier moral example. We would have an answer to the more cynical apes who insisted it was just our 'bad luck' that they have now assumed the more powerful position. I, at least, would like to think that, if my screenplay comes to pass one day, our human offspring will have *some* moral case to make against their own ruthless exploitation at the hands and jaws of other species. If we can do it, why can't they?

15

Why Darwinians Should Support Equal Treatment for Other Great Apes

JAMES RACHELS

James Rachels is professor of philosophy at the University of Alabama at Birmingham. His anthology Moral Problems *was the first modern collection of philosophical essays focusing on practical moral problems, and it has been widely used in undergraduate courses throughout North America. He has written* The Elements of Moral Philosophy *and* The End of Life: Euthanasia and Morality. *His most recent book,* Created from Animals: The Moral Implications of Darwinism, *is an examination of the difference that acceptance of the theory of evolution should make to the way we treat nonhuman animals. Here he discusses the impact that a Darwinian view of the world should have on our relations with the great apes.*

A few years ago I set out to canvass the literature on Charles Darwin. I thought it would be a manageable task, but I soon realised what a naive idea this was. I do not know how many books have been written about him, but there seem to be thousands, and each year more appear.[1] Why are there so many? Part of the answer is, of course, that he was a tremendously important figure in the history of human thought. But as I read the books – or, at least, as many of them as I could – it gradually dawned on me that all this attention is also due to Darwin's personal qualities. He was an immensely likeable man, modest and humane, with a personality that continues to draw people to him even today.

Reflecting on his father's character, Darwin's son Francis wrote that 'The two subjects which moved my father perhaps more strongly than any others were cruelty to animals and slavery. His detestation of both was intense, and his indignation was overpowering in case of any levity or want of feeling on these matters.'[2] Darwin's strong feelings about slavery are expressed in many of his writings, most notably in the *Journal of Researches*, in which he recorded his adventures on the *Beagle* voyage. His comments there are among the most moving in abolitionist literature. But it was his feelings about animals that impressed his contemporaries most vividly. Numerous anecdotes show him remonstrating with cab-drivers who whipped their horses too smartly, solicitously caring for his own animals and forbidding the discussion of vivisection in his home.[3] At the height of his fame he wrote an article for a popular magazine condemning the infamous leg-hold trap in terms that would not seem out of place in an animal-rights magazine today.

For the most part, however, Darwin avoided moralising in his scientific books. Earlier students of nature had viewed the natural order as a kind of moral laboratory in which God's design was everywhere evident, and so they found all manner of moral lessons there. Darwin believed it is a mistake to think about nature in this way. Nature is 'red in tooth and claw'. Rather than embodying some great moral design, nature operates by eliminating the unfit in ways that are often cruel and that do not conform to any human sense of right.

Nevertheless, Darwin did think that something can be learned about morality from the scientific study of human origins. The third chapter of his great work *The Descent of Man* is an extended essay on morality, 'approached', as he put it, 'from the side of natural history'.[4] In that chapter Darwin discusses, among other things, the nature of morality, its biological basis, the extent of our moral duties, and the prospects for moral progress. It is the work of a moral visionary as well as a man of science.

Darwin's remarks about moral progress are especially striking. We are moral beings because nature has provided us with 'social instincts' that cause us to care about others. (The social instincts are, of course, produced by natural selection, as are almost all our traits.) At first, though, the reach of the social instincts does not extend very far – we care only about our near kin and those whom we can expect to help us in return. Moral progress occurs over time as the social instincts are extended ever more widely, and we come to care about the welfare of more and more of our fellow beings. The highest level of morality is reached when the rights of all creatures, regardless of race, intelligence, or even species, are respected equally:

[T]he social instincts which no doubt were acquired by man, as by the lower animals, for the good of the community, will from the first have given to him some wish to aid his fellows, and some feeling of sympathy. Such impulses will have served him at a very early period as a rude rule of right and wrong. But as man gradually advanced in intellectual power and was enabled to trace the more remote consequences of his actions; as he acquired sufficient knowledge to reject baneful customs and superstitions;

as he regarded more and more not only the welfare but the happiness of his fellow-men; as from habit, following on beneficial experience, instruction, and example, his sympathies became more tender and widely diffused, so as to extend to the men of all races, to the imbecile, the maimed, and other useless members of society, and finally to the lower animals, – so would the standard of his morality rise higher and higher.[5]

The virtue of sympathy for the lower animals is 'one of the noblest with which man is endowed'.[6] It comes last in the progression because it requires the greatest advancement in thought and reflection.

What are we to make of this? One possibility is to say that Darwin's moral attitudes were separate from, and independent of, his strictly scientific achievement. In opposing slavery, he was properly sympathetic to one of the great moral movements of his times. In opposing cruelty to animals, he showed himself to be kind-hearted, as we all should be. But no more should be made of it than that. On this way of thinking, the moral views expressed in *The Descent of Man* are just extra baggage, having no more to do with the theory of natural selection than Einstein's reflections on war and peace had to do with special relativity. Like other nineteenth-century writers, Darwin could not resist presenting his thoughts about ethics alongside his scientific work. But we, at least, should keep the two things separate.

There is, however, another way of thinking about Darwin's life and work. Perhaps his scientific work and his moral views were connected, as he apparently believed they were, in a significant way. If so, then we may have reason to view them as one piece, and it may not be so easy to embrace the one without the other. Asa Gray, the Harvard botanist who was Darwin's leading defender in America, took this view. Speaking before the theological faculty at Yale in 1880, Gray declared that

> We are sharers not only of animal but of vegetable life, sharers with the higher brute animals in common instincts and feelings and affections. It seems to me that there is a sort of meanness in the wish to ignore the tie. I fancy that human beings may be more humane when they realize that, as their dependent associates live a life in which man has a share, so they have rights which man is bound to respect.[7]

Asa Gray had identified the essential point. Darwin had shown that all life is related: we are kin to the apes. If this is true, then if we have rights, would it not follow that they have rights as well?

Let me try to explain this point in a little more detail. A fundamental moral principle, which was first formulated by Aristotle, is that like cases should be treated alike. I take this to mean that individuals are to be treated in the same way *unless there is a relevant difference between them*. Thus if you want to treat one person one way, and another person a different way, you must be able to point to some difference between them that justifies treating them differently. Where there are no relevant differences, they must be treated alike.

Aristotle's principle applies to our treatment of nonhumans as well as to our treatment of humans. Before Darwin, however, it was generally believed that the differences between humans and nonhumans are so great that we are almost always justified in treating humans differently. Humans were thought to be set apart from the rest of creation. They were said to be uniquely rational beings, made in God's image, with immortal souls, and so they were different in kind from mere animals. It is this picture of humankind that Darwin destroyed. In its place he substituted a picture of humans as sharing a common heritage, and common characteristics, with other animals.

If we take the Darwinian picture seriously, it follows that we must revise our view about how animals may be treated. It does not follow that we must treat all animals as the equals of humans, for there may still be differences between humans and some animals that justify a difference in moral status. It would make no sense, for example, to argue that clams should be given the right to live freely, because they lack the capacity for free action. Or perhaps the members of some species, such as insects, lack even the capacity for feeling pain, so that it would be meaningless to object to 'torturing' them. Other examples of this type may come easily to mind.

Nevertheless, when we turn to the 'higher' animals, such as the great apes, it is the similarities and not the differences between them and us that are so striking. These similarities are so widespread and so profound that often there will be no relevant differences that could justify a difference in treatment. Darwin argued that such animals are intelligent and sociable and that they even possess a rudimentary moral sense. In addition, he said, they experience anxiety, grief, dejection, despair, joy, devotion, ill-temper, patience, and a host of other 'human' feelings. Ethological studies since Darwin's day have confirmed this picture of them. The moral consequence is that if they have such capacities, then there is no rational basis for denying basic moral rights to them, at least if we wish to continue claiming those rights for ourselves. Chief among those rights are the right to life, the right to live freely and the right not to be caused unnecessary suffering.

It would be easy to overstate this conclusion and to misrepresent its basis. The conclusion is not that the great apes should be granted *all* the rights of humans or that there are *no* important differences between

them and us. There may still be some human rights that have no
analogues for the apes. In an enlightened society, for example, humans
are granted the right to higher education. Because reading is essential for
acquiring such an education, and not even the most intelligent nonhumans can read, it makes no sense to insist that they be given this right.
But the right to live freely and the other basic rights mentioned above do
not depend on the ability to read or on any other comparable intellectual achievement, and so such abilities are not relevant to eligibility for
those rights. Aristotle's principle requires equal treatment where, but
only where, there are no relevant differences.

Partisans of the animal rights movement sometimes represent such
conclusions as based on the genetic similarity between humans and
other apes. But the importance of this fact is easily misunderstood.
Shared DNA is further proof of our kinship with other animals, it
confirms the Darwinian picture, but it is not the bare fact that we share
genetic material with the chimps that forces the moral conclusion. What
forces the moral conclusion is that the chimps, and other great apes, are
intelligent and have social and emotional lives similar to our own. Genes
are important, to them and to us, only because they make those lives
possible.

Before Darwin, the essential moral equality of the great apes – a
category that, of course, includes us as well as the chimps, gorillas and
orang-utans – would have been a surprising claim, difficult to defend.
But after Darwin, it is no more than we should expect, if we think
carefully about what he taught us. Every educated person has now
learned Darwin's lesson about the origins of human life and its connection with nonhuman life. What remains is that we take its moral
implications equally seriously. Darwin himself was optimistic:

> Looking to future generations, there is no cause to fear that the
> social instincts will grow weaker, and we may expect that virtuous
> habits will grow stronger, becoming perhaps fixed by inheritance.
> In this case the struggle between our higher and lower impulses
> will be less severe, and virtue will be triumphant.[8]

Notes

1. My own contribution to the deluge is James Rachels, *Created from Animals:
   The Moral Implications of Darwinism* (Oxford University Press, Oxford,
   1990), which provides a more detailed account of the matters discussed in
   this chapter.

5157

Why Darwinians Should Support Equal Treatment 157

2. This statement, from an unpublished reminiscence in the Cambridge University Library's collection of Darwiniana, is quoted in Ronald W. Clark, *The Survival of Charles Darwin* (Random House, New York, 1984), p. 76.

3. Darwin did, however, defend the practice of vivisection 'for real investigations on physiology'. For details concerning his ambiguous attitude on this subject, see Rachels, *Created from Animals*, pp. 212–16.

4. Charles Darwin, *The Descent of Man, and Selection in Relation to Sex* (John Murray, London, 1871), p. 71.

5. Ibid., p. 103.

6. Ibid., p. 101.

7. Asa Gray, *Natural Science and Religion: Two Lectures Delivered to the Theological School of Yale College* (Charles Scribner's Sons, New York, 1880), p. 54.

8. Darwin, *Descent of Man*, p. 104.

16

Profoundly Intellectually Disabled Humans and the Great Apes: A Comparison

CHRISTOPH ANSTÖTZ

Christoph Anstötz is professor of special education at the University of Dortmund, where he teaches education for the intellectually disabled. He is the author of Grundriss der Geistigbehindertenpädagogik *(Outline of Education for the Intellectually Disabled) and of* Ethik und Behinderung: Ein Beitrag zur Ethik der Sonderpädagogik aus empirisch-rationaler Perspective *(Ethics and Handicap: An essay on the ethics of special education from an empirical-rational perspective). The essay that follows breaks a taboo in the field of special education: it explicitly compares the abilities of profoundly mentally handicapped human beings with those of nonhuman primates. The ethical import of the comparison is, of course, not intended to undermine the excellent progress made in extending equality to the intellectually disabled, but rather to suggest that the basis for this progress points inescapably to a further step.*

Equality and Intellectually Disabled People

If we make a serious effort to understand the idea of equality we find that even in our modern, enlightened world some consequences of this principle are overlooked. In this chapter I argue that the development of this idea, which did not reach intellectually disabled people until the early 1970s (when two declarations of the United Nations were passed) is still not complete. I shall try to show that opposition to discrimination against intellectually disabled people is based on principles that lead to opposition to discrimination against other sentient beings who are also unable to defend their own interests.

The history of this idea of equality can be seen as a story of the development of the moral requirement to give up unjustified forms of discrimination. The great variety of living beings on our planet has given rise to many possible forms of discrimination by beings who could take advantage of these differences. Article 2 of the 1948 United Nations Declaration of Human Rights refers to a number of past forms of discrimination. It rejects 'distinctions of any kind, such as race, colour, sex, language, religion, political or other opinion, national or social origin, property, birth or other status'.[1] Although it cannot be denied that all over the world people violate the idea of equality, we cannot ignore the progress of this idea. As evidence of this, one could point to recently passed German regulations regarding the appointment of women to public offices, or to the most recent developments in South Africa in which the policy of apartheid has finally been abandoned.

The idea of equality is constantly being refined. One hundred and fifty years ago, Harriet Taylor Mill described the essence of this idea in an article about the enfranchisement of women: 'It is an acknowledged dictate of justice to make no degrading distinctions without necessity. In all things the presumption ought to be on the side of equality.'[2] In the second part of the present century the idea of equality gained another important victory on behalf of a so far neglected minority. This victory was also significant for its practical consequences. In contrast to women or racial minorities, this minority is not always able to stand up for its rights. The fourth congress of the International League of Societies for Persons with Mental Handicap (ILSMH) met in October 1968 in Jerusalem and agreed to a 'Declaration on the General and Special Rights of Mentally Retarded Persons'. Representatives from thirty-four countries participated in this decision. The declaration was taken up, discussed and finally accepted by the United Nations. The General Assembly unanimously accepted Resolution No. 2856 (XXVI): 'Declaration on the Rights of Mentally Retarded Persons'.[3] Article 1 expresses the principal basis of the declaration, namely that 'the mentally retarded person has . . . the same rights as other human beings'. Articles 2 to 7 state rights to medical care and educational training, to live in one's own family and to participate in community living. They assert a right to protection against exploitation and disrespectful treatment. Where restrictions of these rights might be necessary, the last article insists on legal safeguards against any kind of abuse. Five years later, these rights for intellectually disabled people were strengthened and supplemented by another UN Declaration: the 'Declaration on the Rights of Disabled Persons'.[4]

These international efforts command worldwide respect. They were accompanied by national exertions in the same direction. In the United States and many other Western countries attempts have been made to remove social conditions that impede the application of the idea of equality to the intellectually disabled. The pioneers of this movement to end discrimination came mainly from Scandinavia. By the early 1960s the Dane Bank-Mikkelsen and the Swede Nirje had gained considerable notice in politics and special education because of their insistence on arranging the living conditions of people with mental disabilities in as normal a way as possible. In the United States, the champion of integration Wolf Wolfensberger took up this idea and discussed it in many articles. This internationally established principle of normalisation is in some way used as a normative test in special education as well as in social politics in order to assess the living conditions of mentally disabled citizens.

With the support of the United Nations Declarations, the move to bring people with mental disabilities into the community of equals gained more and more success. For more than twenty years there has been a trend towards deinstitutionalisation. The aim is to break up the large institutions for mentally disabled people. Living in the community has become part of a whole process of social integration, which is being extended to all aspects of social life. This includes school education, too. Article 2 of the Declaration on the Rights of Mentally Retarded Persons asserts a right to 'such education, training, rehabilitation and guidance as will enable [the mentally retarded person] to develop his ability and maximum potential'. This right is already being realised in several countries. In Germany, teachers for special schools are trained at universities. That means that pedagogical staff must be qualified at a high level in order to give students with mental disabilities a good education and the optimum preparation for an independent and pleasant life in their community.

Though there is no reason to be content with the present situation, the conceptual and actual progress made over the last thirty years in applying the idea of equality in favour of people with mental disabilities cannot be denied.

The Search for the Humanum

During the 1960s and 1970s more and more special schools for people with mental disabilities were founded. The students had to meet certain minimal requirements. The first official guidelines for those schools in Nordrhein-Westfalen, one of the provinces of West Germany, contain the following criteria for entrance.

(a) a mental and psychological standard of development, that allows for the recognition of objects and enables the disabled to

manipulate objects in a purposeful way and spend several minutes in appropriate activities;
(b) the ability to understand simple verbal or gestural information;
(c) the ability to learn social competence.[5]

After school education is finished, so the official guidelines noted, the students should be able to take over 'tasks in the household, kitchen and garden', to have an appropriate 'contact with animals', to use knowledge in 'health care and domestic care of the sick', to 'handle tools and machines' and possess some 'industrial skills'.[6] Regarding this group of people, there seemed to be hardly any problems in justifying educational training. On the one hand, of course, one cannot equate the standard of schools for people with mental disabilities with standards of other state schools. But on the other hand, the educational goals involve skills and abilities, which are not of the kind to evoke serious opposition from those who have little or no contact with people with mental disabilities. The people concerned are adolescents or adults who are mentally disabled and therefore impaired in other aspects of their personality, but who will, when educated, be able to be independent in much of their everyday lives. To apply Article 1 of the UN Declaration of Human Rights in this case seems straightforward: 'All human beings are born free and equal in dignity and rights. They are endowed with reason and conscience and should act towards one another in a spirit of brotherhood.'

To found schools for people with mental disabilities and thereby give expression to our social concern for these members of society was only a first step. The conviction increased that mental disability is never a reason for withdrawing the right to an education in school. This position had consequences. If it is agreed that a mental disability cannot be a reason to deny a member of society the right to education in school, the criteria for entrance cited above can no longer be accepted. Instead it seemed obvious that the idea of equality demanded the inclusion of all students of school age regardless of their degree of mental disability. Hence the criteria of entrance cited above were dropped from the second official guidelines for schools for people with mental disabilities, drawn up in 1980. Shortly before, a ministerial prescription[7] had ordered the admission of profoundly mentally disabled people, whose educational integration in school had not previously been planned.

In the relevant literature, those students are often described as not developing beyond the abilities of an infant of a few months old. The new official guidelines of Nordrhein-Westfalen, which were developed with the assistance of experts in the field of special education, express the view that even this low level will not be reached by every student:

Learning behaviour is characterised by impairment of the reception, preparation and storing of information and of the mode of expression. The variety of learning behaviour may range from:
- as yet, no observable interest in learning, even concerning vital needs, to an interest in learning regarding vital needs . . .
- as yet, no recognisable ability to respond to personal expressions and reactions, to the ability to respond to situations and objects . . .
- as yet, no perceptible reactions to feelings and stimuli, to a mainly action-oriented learning . . .
- as yet, no observable communication, to a reduced ability of receiving, processing and expressing verbal information.[8]

There are a lot of other deviations, such as:

- little developed ability to co-ordinate sensorial impressions and adequate movements and behaviour . . .
- missing or reduced ability to react to persons and situations or to distance oneself from them.[9]

Against the background of these characterisations it is worth looking again at Article 1 of the Declaration of Human Rights: 'All human beings are born free and equal in dignity and rights. They are endowed with reason and conscience and should act toward one another in a spirit of brotherhood.'

Whereas it was plausible to apply this article to the first group of people with mental disabilities entering the special schools, one must have severe doubts about its application to the second group of more severely mentally disabled students. How can we accept that a student is gifted with 'reason and conscience', when that student does not respond to any stimuli in a perceptible way, is unable to take part in communication, and cannot react to other people or his or her surroundings at all?

To a certain degree these doubts can be resisted. One may reply that, first of all, the results of an intensive education as well as integrated living conditions must be seen before one judges the capacities of a profoundly mentally disabled person. The official guidelines consider this aspect by talking about 'as yet' no perceptible responses or 'as yet' no perceptible participation in communication. Secondly, even if after many years of educational efforts under optimal living conditions there is still no change, one can always point to the general progress in knowledge which leads to a continuous improvement in our educational abilities. This reply is well-founded, but it is also one-sided. It is neither logically nor empirically defensible to consider only the limitations of educational programmes and not the limitations of students.

One cannot deny that a profoundly mentally disabled young man, let us say at the age of eighteen, may never exceed the minimal capacity of an infant, despite committed educational efforts and optimum social surroundings. To deny this would mean closing our eyes before an unpleasant reality and giving our wishes the status of pedagogical premises which defy human experiences.[10]

But it is not necessary to look at the lowest end of human capacity in order to analyse the moral dilemma which is associated with the inclusion of profoundly mentally disabled students among the clientele of special education. The fact that these adolescents seem to be missing – perhaps for all their lives – what we commonly consider characteristic human abilities has indeed evoked a lot of confusion in the field of special education, and especially in its ethical aspects. Faced with this new situation and confronted with the extremely minimal capacities of these human beings, we are forced to look again for what should be regarded henceforth as human, as the *humanum*.

In the context of a research project, Pfeffer tells of the impression of one of his colleagues who makes contact with a profoundly mentally disabled student for the first time:

> Everything that makes a difference between humans and animals seems to be missing: abstract thinking and language as typical human attributes could be eliminated right away. What remains, then? . . . What is a human being, this measure of everything, if the profoundly mentally handicapped are humans too?[11]

Profoundly Mentally Disabled Humans and Nonhuman Primates: A Comparison

For the first time in special education humans were compared to animals. Such comparisons were regarded as an especially convincing strategy for upholding a special moral status for humans among all other living beings, even in cases of profound mental disability. 'In biological anthropology', the well-known Dutch educationalist Stolk tells us, 'you try to find out this typical humanum by comparing humans to animals'.[12] Parents of severely mentally disabled children also draw such comparisons, as an example from the same source demonstrates:

> The parents of Johan show the photo album with their son. The photos of successive years indicate increasingly how serious the disability of their son is. Then the photos stop. The boy is now 14 years old, but there is no photo since he was 7. Those who know

Johan don't ask why. His father says of him: 'You get more love from a pet.' After a long period of silence, he adds with bitterness: 'He lives like a plant, like a poorly blooming weed.' And after another long pause: 'But the crazy thing is, he always remains your child.'[13]

From this example Stolk draws the following conclusion:

If we compare human beings to animals, the mentally disabled seem to lack those attributes which are commonly described as typically human. In some respects profoundly mentally disabled people even fail to bear comparison with animals.[14]

What does 'in some respect' exactly mean? Andreas Froehlich is a special educationalist who has done a lot of research in the field of advancement of severely mentally disabled people.[15] One of his most recent articles about communication with profoundly mentally disabled people starts with an attempt to state the existential meaning of communication. He notes that in medical guidelines on the limits of medical treatment for profoundly disabled newborns it is pointed out that the ability to communicate may become the ultimate measure of human individuality. Then he writes: 'Life and ability to communicate virtually go together, and thus this communication gets more and more importance for our human self-understanding.'[16] Together with Ursula Haupt, a psychologist, he tries to develop a system of communication in which perception, emotions, cognition, movement, body experience and social experience all interact with each other and with communication. He emphasises the interdependences and influences of the separate areas on communication. Against the background of a profound mental disability, different aspects of communication were treated separately: visual and tactile, vibratory, smell and taste. Froehlich further differentiates the physical area into body contact, proximity, orientation, gaze, eyebrows, posture, facial expression and intonation. He also describes, from the viewpoint of educational advancement, the influence of feelings, cognition, and social and physical experiences.

It is not quite clear what Froehlich means when he says that he has found a 'preliminary possibility of describing complex human development'.[17] If this is supposed to mean that his scheme is only applicable to human beings, it clashes with what we know about, for example, nonhuman primates. Jane Goodall's book *The Chimpanzees of Gombe*[18] describes the results of twenty-five years of research on chimpanzees. In the sixth chapter, on communication, Goodall begins by describing how chimpanzees communicate feelings such as fear, stress, anger, pleasure and so on. Different ways of communication are analysed, including visual communication, tactile and auditory communication, olfactory communication, and various combinations of these.

It is revealing to see that every aspect of communication discussed in Froehlich's article on communication with severely mentally disabled people is also discussed in Goodall's chapter on the communication of chimpanzees. But that is not all. There is nothing that humans with the most serious intellectual disabilities can do or feel that chimpanzees or gorillas cannot; moreover, there is much that a chimpanzee or a gorilla can do that a profoundly mentally disabled human cannot do. This includes the characteristics generally regarded as distinctive of human beings. To show this, here are some examples relevant to language, intelligence and emotional life.

Like other educationalists in this field Froehlich points out that profoundly mentally disabled people have as little real linguistic competence as very young children. Compare this with Francine Patterson's descriptions of the abilities of two gorillas, Koko and Michael, in Chapter 6 of this book and elsewhere.[19] Both Koko and Michael take the initiative in beginning conversations with humans. They use their vocabulary in creative and original combinations to describe their surroundings, their feelings, their wishes and their personal experiences. They understand spoken English and have been taught to read some written language. Patterson adds that dialogues with gorillas are like dialogues with small children and therefore in many cases special interpretations and completions are necessary. Jane Goodall's research mentioned above reports on many successful experiments teaching chimpanzees and other nonhuman primates to use symbolic means of communication. Allen and Beatrice Gardner taught Washoe to use American Sign Language, as described by Roger and Deborah Fouts in Chapter 4. At the age of 5 years, she understood 350 different symbols and she was able to use about 150 of them in an active way. She transfers signs from one context to another in a correct manner. Thus she learned the meaning of 'open' in connection with doors and used it correctly to refer to opening the refrigerator or other containers. The scientists Duane M. Rumbaugh and Timothy V. Gill found during their lessons with the chimpanzee Lana that she correctly uses stock sentences in various contexts.[20] Like Francine Patterson they observed that Lana initiated conversations and invented new variations and combinations of known words. Lana even created new names for certain objects by freely combining learned signs.

Even without further examples and deep analysis it can be said that these apes show a linguistic competence that cannot be achieved by profoundly mentally disabled humans even after long and intensive training. At the end of their study with Lana, Rumbaugh and Gill also mentioned the fundamental dependence of linguistic skills on intelligence: 'We believe that the success Lana has had so far in acquiring linguistic-type skills supports our view of language – that the foundations of language are to be found in the processes of intelligence.'[21] What do we know about the higher cognitive abilities of chimpanzees and other nonhuman primates as revealed, not only in the use of language but also in solving other problems?

Research on the mental capacities of chimpanzees and gorillas reaches further back than investigations on severe mental disability in human beings, which has only recently become a topic of educational and psychological research. Most of the currently used intelligence tests are not suitable for profoundly mentally disabled people, because the tests demand a level that is simply too high for these people. But Patterson reports using the Stanford–Binet Children's Intelligence Test with Koko (her results were in the below-average human range). Patterson also refers to other tests she carried out with Koko, including tests of intelligence, of development and of language, such as the Cattel Infant Intelligence Scale, the Bayley Scales of Infant Development, the McCarthy Scales of Children's Abilities, etc. From the results she concludes that there may be much more happening in the minds of gorillas than we commonly suppose today. An experiment by Doehl, cited by Goodall, demonstrates the ability of Julia, a six-year-old chimpanzee, to calculate, in a deliberate manner, how to reach, by means of five separate steps, a box with a banana in it. In the experiment two series of five closed and transparent containers were put before Julia. One of the goal boxes was empty, the other contained a banana. To open the box containing the banana, Julia had to get a distinctively shaped key from another box. But this box was closed too and could only be opened with its own distinctive key. This key Julia had to take out from a third box, also locked and also requiring its own key . . . and so on. Working backwards from her desire to obtain the key for the last box with the banana, Julia was able to choose the right one of the initial two keys, lying in the first two open boxes.[22]

Many other observations by Goodall and her colleagues demonstrate that the mental capacities of nonhuman great apes are far above the level of profoundly mentally disabled humans. This is quite evident when we remember the cognitive abilities referred to in the guidelines mentioned in the previous section and compare them to the abilities of Julia, Lana, Koko, Michael or Washoe.

The scheme for communication devised by Andreas Froehlich and Ursula Haupt takes another important aspect into consideration: the realm of feelings or emotions. Emotions cannot be observed directly. Sometimes in special education they are treated as if they represent a substantial difference between humans and animals. However, as early as 1872, in his *Expression of the Emotions in Man and Animals*, Charles Darwin had shown by detailed observations that the expression of emotions in nonhuman primates is closely analogous to that in human beings. Often he describes in a subtle way how feelings of joy, pleasure, excitement, and also of pain, anger, disappointment and fright are expressed. The systematic observations of modern researchers such as Goodall refute any claims that there is something unique about humans with regard to their expression of feelings and emotions. These observations provide convincing evidence of a rich emotional life in nonhuman great apes. For example, fifteen different calls of the chimpanzees were recorded by sound-spectrograph-analysis. From this basis Goodall and her colleagues identified further variations over more than 50,000 hours of observation. By a controlled rating system they found a definite congruence in more than thirty discrete calls, expressing feelings of fear, confusion, sexual excitement, pleasure, social apprehension and so on. They report similar findings about the emotional meaning of different facial expressions, postures and movements for communication and interaction.

Perhaps these results are still not sufficiently well established. But there is no doubt that they give more definite grounds for accepting that chimpanzees have a rich emotional life than any comparable information about profoundly intellectually disabled people does for our belief in the richness of their emotional life. Statements and presumptions about the inner perspective of the profoundly intellectually disabled, and about their feelings and moods, are presently supported only by the subjective impressions of persons trying to build contacts with them. They often live lost in their own autistic world so that over a longer term, little can be found out about their inner lives, and we must resort to speculation. So, for example, in the context of a research project Pfeffer cited the following observations from his colleagues concerning a profoundly disabled student:

[She] had a really pretty face. . . . But in this sympathetic face there were hardly any changes that might give any indication of an interest in what happened around her. It was as if the girl was in a glass case in a corner of the classroom, cut off from her environment and alone with her body and herself. She showed no reaction to touch or noise, she seemed to see nobody and nothing.[23]

Later on another colleague reported:

Pia was strange and inaccessible to me. I did not understand her and she did not make herself understood by me. The usual means

of contacting strangers, the competence of verbal communication, were totally missing.[24]

Even when it is realised that special training over a long period often
leads to recognisable progress, the limits of education are not to be
denied. Pointing to another student in his project, Pfeffer says:

> Like all the other profoundly mentally disabled people Wolfi, too,
> has great difficulties in communicating . . . The human environ-
> ment is especially unavailing for children with strong autistic
> traits to whom we can apparently make no contact.[25]

Towards a Consistent Idea of Equality

Our considerations started with the statement that people with a mental
disability have come late to the rights laid down in the Declaration of
Human Rights of 1948. If they have completed training in special
schools, many people with mental disabilities today will be able to lead
an independent life in many respects. But in general, a protective
environment and social services are necessary to enable them to live a
satisfying life in their community. We then noted that special education
encountered difficulties when at the end of the 1970s profoundly
mentally disabled children were admitted into schools. The difficulties
were not just in finding the best means of advancing the development of
these children, but also over the moral justification of the rights to
education claimed for them. Distinctively human qualities, easily recognisable in mildly and moderately mentally disabled people, could no
longer provide a decisive moral basis for this right, because among
profoundly mentally disabled people, such qualities are at best found
only in a rudimentary form.

In the normative discipline of special education this situation led
directly to the comparison of humans and animals in order to conserve
the superior moral status of humans among all feeling beings on the
planet. Here appeared two incompatible facts: on the one hand it had to
be recognised that the distinctive human qualities are virtually absent
from people with profound mental disabilities. On the other hand there
was a growth in our knowledge of the existence of so-called typical
human characteristics, such as language, intelligence and emotions, in a
high degree in other nonhuman living beings. As the previously mentioned research shows, these findings do not rest on any sentimental
basis, but on a solid grounding of fact. Knowledge of the varied social

life of nonhuman primates has recently become available to the German-speaking public in a number of popular scientific books.[26]

Nevertheless, in special education the conviction that still predominates is one based on an idea of equality which proceeds from an image of a human being gifted with reason and conscience. Where it is recognised that those typical human qualities are really neither typically human nor present in all human beings, the idea of equality suffers a metamorphosis which contradicts its fundamental sense. Faced with this situation Stolk suggests 'that in giving an answer to the question of humanness we should not start from the differences between human beings but from what is common to them, regardless of their abilities and qualities'.[27] He cites a proposal intended to reveal this humanness, this new *humanum*: 'Human is every being born from a human being.'[28]

Anyone who interprets the idea of equality so as to find the criterion of equality only in membership of the species *Homo sapiens* can include profoundly mentally disabled people in the community of equals. It is done as Stolk describes, 'regardless of their abilities and qualities'. This interpretation is as psychologically strong as it is morally weak. For it prefers all and only members of that group who created this interpretation, discriminating without valid reason against all other living beings who are not members of this privileged community of equals.

The attempt to realise the idea of equality is made more difficult when those who are not members of the relevant moral community are unable to appeal against their exclusion. Not even the most intelligent chimpanzees can protest, either directly or through representatives drawn from their own kind, against the deprivation of liberty, against being used in painful medical experiments, against being killed for food, or against being exhibited in zoos and circuses. On the other hand, according to the Declaration of the United Nations, the profoundly mentally disabled human is protected from any kind of abuse and degradation, merely on the grounds of membership of the species *Homo sapiens*. Should the deeper sense of the idea of equality, on which human rights is based, demand that we provide for the interests and needs of humans but allow discrimination against the interests and needs of nonhuman beings? Wouldn't it be strange if the same idea contains the claim for equality and the permission for discrimination, too? Can this idea really involve sympathy and cruelty at the same time?

In this situation special education is at a moral crossroads. It could stay on its traditional path, content that the progress of the idea of equality finally has reached all its clientele. Now that it is well known, particularly in this discipline, that those typically human qualities can no longer provide a relevant moral foundation for the way in which we treat all human beings, the point of view could be taken, without any consideration of other consequences, that membership of the human species is enough. Perhaps this would happen without any deliberate desire, on the part of members of the discipline, to support the traditional view that in future (for example) painful medical experiments might be performed on beings like Julia but not on those like Pia, simply because Julia is a chimpanzee and Pia is a human being.

The discipline of special education for severely mentally disabled people could choose another way which makes possible a more thorough-going pursuit of the idea of equals. We have to realise that this discipline deals with research and the advancement of minimum needs which require respect even under circumstances in which typically human abilities are no longer found. It is precisely this discipline that has enough moral standing to point out that the fundamental sense of the idea of equality demands principled and equal respect for every need and every interest, no matter whether it is a need of a Julia or of a Pia. Of course it is not the task of special education to care about the destiny of those miserable apes, defencelessly vegetating in laboratories, often until they meet their brutal death. But the discipline offends against its own principles if its ethical disputes about justifying the educational requirements of profoundly mentally disabled humans are settled by the arbitrary exclusion of living beings who are sentient and defenceless, too, and whose only 'fault' is not to be a member of the species *Homo sapiens*.

The knowledge we have today about profoundly mentally disabled humans and nonhuman primates gives strong reason to revise the traditional interpretation of the idea of equality. The time has come to see the community of equals no longer as a closed society, but as an open one. The admission of nonhuman primates and the guarantee of certain fundamental rights in favour of all members of such a community, including profoundly mentally disabled humans, would be a first important step. These rights should include the right to life, the protection of individual liberty and the prohibition of torture.

Notes

1. Universal Declaration of Human Rights (Official Records of the third session of the General Assembly, Part I, 10 December 1948. United Nations, Palais de Chaillot, Paris, 1949), p. 72.
2. H.T. Mill, 'Enfranchisement of women', in J. S. Mill and H. T. Mill, *The Subjection of Women: Enfranchisement of Women* (Virago, London, 1983), S. 9.
3. Declaration on the Rights of Mentally Retarded Persons (Resolution adopted by the General Assembly during its twenty-sixth session, 20

December 1971. General Assembly Official Records Suppl. No. 29 (A/8429). United Nations, New York, 1972), p. 93.

4. Declaration on the Rights of Disabled Persons (Resolution adopted by the General Assembly during its thirtieth session, 9 December 1975. General Assembly Official Records Suppl. No. 34 (A/10034). United Nations, New York, 1976).

5. *Richtlinien für die Schule für Geistigbehinderte (Sonderschule) in Nordrhein-Westfalen* (Schwann, Düsseldorf, 1973), p. 6.

6. Ibid., p. 12.

7. *Aufnahme Schwerstbehinderter in Sonderschulen* (Kulturminister des Landes Nordrhein-Westfalen v. 12.7.1978 – II A5.365/0 – 1831/78).

8. *Richtlinien und Hinweise für den Unterricht. Förderung schwerstbehinderter Schüler* (Greven, Köln, 1985), pp. 6ff.

9. Ibid.

10. J. M. Kauffman and J. Krouse, 'The cult of educability: searching for the substance of things hoped for; the evidence of things not seen', *Analysis and Intervention in Developmental Disabilities*, vol. 1 (1981), pp. 53–60.

11. W. Pfeffer, *Förderung schwer geistig Behinderter. Eine Grundlegung* (Edition Bentheim, Würzburg, 1988), p. 128.

12. J. Stolk, 'Geistig behindert mit dem Verlangen, auch jemand zu sein' in J. Stolk and M.J.A. Egberts (eds), *Über die Würde geistig behinderter Menschen* (Lebenshilfe Selbstverlag, Marburg, 1987), pp. 5–34.

13. Ibid., p. 8.

14. Ibid., p. 10.

15. A. D. Froehlich, 'Ganzheitliche Kommunikationsförderung für schwerer geistig behinderte Menschen', in A.D. Froehlich (ed.), *Lernmöglichkeiten*, 2nd edn (HVA Schindele, Heidelberg, 1989), pp. 17–44.

16. Ibid., p. 17.

17. Ibid., p. 19.

18. J. Goodall, *The Chimpanzees of Gombe. Patterns of Behavior* (The Belknap Press of Harvard University Press, Cambridge, MA, 1986).

19. F. Patterson, 'The mind of the gorilla: conversation and conservation', in K. Benirschke (ed.), *Primates, the Road to Self-sustaining Populations* (Springer, New York, Berlin, Heidelberg, 1986), pp. 933–47.

20. D. M. Rumbaugh and T. V. Gill, 'Lana's acquisition of language skills', in D. M. Rumbaugh (ed.), *Language Learning by a Chimpanzee. The Lana Project* (Academic Press, New York, San Francisco, London, 1977), S. 191, pp. 165–92.

21. Ibid.

22. J. Döhl, 'Über die Fähigkeiten einer Schimpansin, Umwege mit selbständigen Zwischenzielen zu überblicken', *Zeitschrift für Tierpsychologie*, vol. 25 (1968) pp. 89–103.

23. Pfeffer, *Foerderung schwer geistig Behinderter*, p. 111.

24. Ibid., p. 115.

25. Ibid., p. 126.
26. D. Fossey, *Gorillas im Nebel. Mein Leben mit den sanften Riesen* (Kindler, München, 1989); J. Goodall, *Wilde Schimpansen. Verhaltensforschung am Gombe Strom* (Rowohlt, Reinbek b., Hamburg, 1991); F. de Waal, *Wilde Diplomaten. Versoehnung und Entspannungspolitik bei Affen und Menschen* (Hanser, München, Wien, 1991).
27. Stolk, 'Geistig behindert mit dem Verlangen', p. 14.
28. Ibid., p. 15.

17

Who's Like Us?

HETA HÄYRY and MATTI HÄYRY

Heta Häyry is assistant professor of practical philosophy at the University of Helsinki, and the author of several books, among them Animal Welfare *(in Finnish) and* The Limits of Medical Paternalism. *Matti Häyry is junior research fellow in the same department, and the author of a recent book,* Critical Studies in Philosophical Medical Ethics. *In this jointly written contribution they construct and then defend, step by step, a formal moral argument which has as its conclusion the essence of the Declaration on Great Apes: that we should grant equal basic rights to all great apes – human and nonhuman.*

The Declaration on Great Apes states that chimpanzees, gorillas and orang-utans, as well as human beings, should be granted the rights to life, liberty and the absence of deliberately inflicted pain. The argument on which this statement is founded consists of the following premises and conclusions:

P1 Beings who are equal in the moral sense ought to be treated equally.
P2 Beings are equal in the moral sense if their mental capacities and emotional lives are roughly at the same level.
P3 The mental capacities and emotional lives of human beings and other great apes are roughly at the same level.
C1 Therefore, human beings and other great apes ought to be treated equally.
P4 Human beings ought not to be killed, imprisoned or tortured unless certain specific conditions prevail.
C2 Therefore, other great apes ought not to be killed, imprisoned or tortured unless the same specific conditions prevail.

We believe that this argument is essentially sound, at least in the sense that those human beings who imprison, torture and kill chimpanzees, gorillas and orang-utans for scientific (or allegedly scientific) or commercial purposes are acting wrongly and should be stopped. (We say 'at least' since one might feel uneasy about bringing human law and order to the orang-utans, gorillas and chimpanzees themselves, i.e. to intervene coercively if orang-utans, gorillas and chimpanzees within their own natural communities act violently towards each other. Cultural toleration can perhaps be argued for by referring to the probable harm that human interventions would inflict on individuals within these communities.)

But what about those scientists, philosophers and lay persons who disagree with us? Obviously, they must claim that one of the premises of the argument is false, or that the logical inference from the premises to the conclusion is somehow invalid. Let us see how such claims could be defended.

Equal Treatment for Equals

In the first premise of the argument (P1) the words 'equal' and 'equally' should not be taken to mean 'similar' and 'similarly', as those opposing equality between races, species and the sexes frequently suggest. Beings who are equal in the moral sense need not be exactly like one another. The obvious individuality of male adult human beings, for instance, has never deterred those who have declared that all men should be treated equally. The idea is, rather, that there are certain relevantly similar features in all beings who belong to the same category of moral equals which make them members of the category.[1]

The same point applies, with necessary changes, to the equal treatment of beings who are relevantly similar. Consider certain advanced medical procedures performed on humans. At any given moment, for example, there are a number of people all around the world who ought to be given blood transfusions. But even though all people are presumably equal in the moral sense, it does not follow that blood transfusions ought to be given to everybody. There are those who do not need additional blood at the present, and there are those who do not wish alien blood to circulate in their veins. Equal treatment in this case does not mean that perfectly healthy individuals should be seized in the street and provided with blood transfusions. Nor does it mean that blood or blood products should be forcibly injected into the veins of competent adults who, due to their religious beliefs, oppose the idea. Equal treatment implies that every being in a category of moral equals ought to be treated according to its or her or his needs and desires. Some theorists have called this type of equality 'equality of consideration', stating that the interests of all members of the community of equals should be taken equally into account.[2]

Given these qualifications, P1 can hardly be attacked from any quarter. Almost all substantial theories of justice and equality can be founded on the general principle that relevantly similar beings ought to be treated in a relevantly similar manner. The principle in this form is a tautology rather than an action-guiding moral statement, and additional premises are therefore needed to give the argument its normative content.

The Community of Equals

The second premise of the argument (P2) states the criteria which divides beings into different moral categories. According to the premise, communities of moral equals consist of beings who are approximately at the same level as regards their mental capacities and emotional lives. For instance, assuming that trees have no thoughts or feelings or social contacts with each other, they belong to the same moral category as stones and raindrops. (The assumption can be false in some sense, but this does not influence the following argument, which is hypothetical.) This similarity means that if one can justifiably kick a stone or capture a raindrop, one can also legitimately kick trees and build fences around them. Irrelevant differences, like the fact that trees are living organisms while stones and raindrops are inanimate objects, do not count in the matter.[3]

There are a number of competing views which state that psychological and social abilities are not the proper basis for assessing how different beings ought to be treated. The alternative criteria which have been suggested range from species membership, through merit and the ability to claim one's due, to general utility. We shall argue below that all these alternative solutions are either unsound or yield the same normative conclusions as the original premise on which the Declaration is founded.

But whatever the status of the Declaration's approach to equality, the third premise of the argument (P3) can also be independently challenged. Many scientists and philosophers have repeatedly argued that the mental capacities and emotional lives of chimpanzees, gorillas and orang-utans are not sufficiently developed to justify the comparison with human cognition and feelings. The falsification or verification of such claims is, of course, mainly a matter for empirical scientists. There are, however, two conceptual points which seem to support the view expressed in the Declaration. First, most scientists who maintain that human beings and other great apes are too unlike each other to be counted as moral equals also hold the view that experimentation on chimpanzees, due to the striking similarities between the species, is the only way to ensure that new medical procedures are safe for humans. Second, although it may be true that adult chimpanzees, gorillas and orang-utans are, as a rule, less intelligent and sensitive than average adult human beings, it should be remembered that the same remark also applies to many humans, including very young children and people with a mental handicap. Assuming that the general principle stated in premise P2 is valid, it is difficult to refute the argument by attacking its third premise (P3).

Basic Moral Principles

The fourth premise of the argument states the qualified moral principle that human beings ought not to be killed, imprisoned or tortured unless certain specific conditions prevail. By these 'specific conditions' different theorists have meant slightly different things. In the Declaration it is stated that killing human beings may be justifiable in self-defence, and that their imprisonment can be legitimate only if it is the result of adequate legal processes. According to the Declaration, an adequate legal process should not lead to the imprisonment of human beings (or other great apes) unless they have been convicted of a crime, or their continued detention is in their own best interest, or they constitute a threat to the safety of others. The deliberate infliction of severe pain is rejected in the Declaration without exceptions.

All these qualifications can in fact be contested. As for the clause on taking lives, there are those who believe that killing is always wrong and should not be sanctioned even in self-defence. There are also those who assert that actions which are condemned as criminal in modern societies do not justify imprisonment. Crime, they hold, is created and maintained by society itself, and the rules of crime prevention are defined so as to benefit the rich and the powerful and to oppress the poor and the powerless. Besides, they argue, confinement neither cures criminality nor protects the safety of ordinary citizens, since prisons promote violent and antisocial behaviour instead of eradicating it.

In addition to these pacifist and abolitionist objections, the qualifications stated in the Declaration can also be challenged from a more self-centred point of view. Not everybody believes that the individual's own best interest can be served by restricting his or her liberty. And some of us may find it theoretically unsound to condemn the infliction of pain under all circumstances while condoning, at the same time, killing in self-defence.

But although there may be differences of opinion when it comes to applying the moral principles stated in the Declaration, this does not in any way undermine the validity of those principles themselves. Details aside, most people surely believe that human beings ought not to be killed, imprisoned or tortured unless there are some exceptionally weighty reasons for doing so. Consequently, assuming that human beings and other great apes are to be treated equally, as the first conclusion of the argument (C1) affirms, chimpanzees, gorillas and orang-utans should be similarly protected against death, constraint and deliberately inflicted pain. The preferential treatment of one species at the expense of the other three can only be justified by showing that C1 is invalid. And since the first premise of the argument was found to be noncontroversial and the third premise beyond conceptual proof, the only way in which philosophers can argue that human beings ought to be treated differently from other great apes is to attack the second premise (P2).

Against the Mental Criteria of Moral Equality

There are two ways in which the psychological and social criteria for moral equality are usually interpreted, and both interpretations introduce their own difficulties. First, it is possible to define the sufficient 'mental capacities and emotional life' referred to in the Declaration strictly, so as to ensure that beings who fulfil the criteria cannot fail to deserve equal consideration and treatment. Seen from the viewpoint of adult human beings, it would be natural to include, for instance, vivid self-awareness and mutual verbal communication among the necessary requirements. The problem with this solution is, however, that it would exclude human infants and intellectually disabled human beings as well as most other animals from the community of moral equals. Many people seem to feel that this would be an unacceptable conclusion, especially as regards the human beings who do not meet the standards.

Second, it is possible to define the criteria broadly, and state that even rudimentary self-awareness, combined with the ability to suffer pain and distress, is sufficient for membership in the community of equals. This, manifestly, is the view taken in the Declaration. Those who criticise the mental approach can argue, however, that the solution would extend the community of equals beyond all reasonable limits. If the ability to suffer is used as the decisive criterion for equal treatment, then most 'higher' animals – dolphins, pigs and the rest – would have to be granted the rights to life, liberty and the absence of deliberately inflicted pain. And such a result would, according to the opponents of the mental approach, be patently absurd.

This challenge can be met in two ways without rejecting the mentalistic view. On the one hand, the extension of the community of equals beyond great apes does not necessarily deter those who find the Declaration defensible in the first place. Although further expansions in the category of equals would lead to drastic changes in the views and lifestyles of humans, this does not prove that the expansions would be absurd. The abolition of slavery presumably altered the way of life in the Southern states of the USA, but few people today would regard this as an argument for slave-holding. The Declaration does not rule out the possibility that other animals besides great apes may have to be granted equal rights.

On the other hand, it is also possible to meet the challenge by analysing in more detail the relationship between the rights and mental abilities of different beings. There are animals, including great apes like ourselves, who, in addition to their ability to suffer from constraint and physical pain, are aware of themselves as distinct entities whose existence is temporally continuous. Only these individuals can suffer from their own demise, or the thought of their own nonbeing in the future, and only they possess the right to life in the strict sense. There are also animals who lack self-awareness but who are sentient and capable of being distressed if they are imprisoned – they have the rights to liberty and the avoidance of torture. Finally, beings who are merely sentient, i.e. only sensitive to physical suffering, are entitled to protection against deliberately inflicted pain.

The definition of these three subcategories of moral equals is, ultimately, an empirical issue. Even without further empirical scrutiny, however, the division shows that whatever difficulties there may be within the mentalistic view, these do not constitute compelling reasons for rejecting it. This is not to say that the approach is universally accepted. But unless an alternative theory can be found which clearly supersedes the psychological criteria, the foundation of the Declaration cannot be effectively criticised.

Membership in a Species

One of the most popular alternatives to the mentalistic approach is the employment of biological criteria. At the core of the view is the claim that the community of moral equals can be extended only to human beings, i.e. to the members of the species *Homo sapiens*. When the theory is presented in its orthodox form, no reference is made to the abilities, achievements or opinions of individual human beings. What is considered focal is that humankind as a whole is entitled to preferential treatment among the species.

The view that a species is morally important can be defended by two methods, neither of which leads to tenable results. First, some proponents of the view argue that it is God's will that human beings use animals to satisfy their needs. Second, others argue that the creation of law and society sets human beings apart from other animals. It would be absurd to burden chimpanzees, gorillas and orang-utans with legal or social duties, and it would therefore also be absurd to provide them with the corresponding rights.

Unfortunately for theistic speciesism, the Bible does not in fact give any unequivocal grounds for killing, imprisoning or torturing non-human animals.[4] What Genesis claims is that human beings ought to be the rulers of Creation, not that they should use other animals as a means to their own ends. It is not, after all, part of the ruler's duties to eat his or her subjects nor imprison them without adequate grounds. Even assuming that Judeo-Christian beliefs are relevant to critical morality, the Declaration cannot be attacked by appeals to God's will.

The humanistic view, in its turn, is based on facts but it falls short of applying these facts consistently. Chimpanzees, gorillas and orang-utans have not participated in the making of prevailing laws and social policies, but the same observation applies to the majority of humans. The 'mankind' which is responsible for most modern regulations and political procedures consists mostly of white male adult humans, not of all human beings without distinction of age, race, sex or social position. Subsequently, the humanist should not draw the boundary of moral equality between human and nonhuman beings, but between white male adult humans and other human and nonhuman animals.

Merit

Other criteria for determining how different beings ought to be treated include merit, the ability to claim one's due and general utility. None of these approaches fares, however, any better than pure speciesism.

Those moralists who believe that theories of justice and equality should centre on the concept of merit state that all beings, regardless of their natural features or social background, should be treated according to their merits, or should be 'given their just deserts'. Hardworking people and productive domestic animals deserve the full respect and consideration of others, whereas beings who do not earn their dues deserve considerably less. When this view is applied to laboratory animals, including chimpanzees, gorillas and orang-utans, the moral is two-fold. On the one hand, the beasts who 'serve well' as experimental subjects also deserve to be treated well by their examiners. On the other hand, however, laboratory animals are by definition exposed to many hazards and indignities. Chimpanzees, gorillas and orang-utans who are supposed to further scientific progress often live their entire lives in captivity, and they are frequently subjected to physical pain. Moreover, even if senseless cruelty towards these human-like apes may be rare, their lives are in many cases shortened and abruptly terminated as a result of the experiments.

Even those who do not find it disquieting to think about imprisoned chimpanzees, gorillas and orang-utans have to admit, however, that the theory of merits and deserts is, in the last analysis, untenable in the present context. There are many human beings who, like other great apes, are unable to earn their dues by productive work as understood by the able-bodied and able-minded section of the human community. Yet most people would, and rightly, hesitate to cast these unfortunate individuals in the role of experimental animals. When it comes to members of our own biological species, we tend to complement the ethics of deserts with the ethics of compassion or rights. But the theory of just deserts gives no adequate grounds for limiting the application of these moral safeguards to humans. The theory assumes rather than argues that chimpanzees, gorillas and orang-utans should be treated differently from the other great ape.

The Ability to Claim One's Due

One argument which is frequently employed to defend speciesism is based on the fact that chimpanzees, gorillas and orang-utans do not have the ability to claim their dues, whatever the criteria for measuring those dues may be. The proponents of the argument seem to presume that beings can only have rights if they have themselves formulated these rights, and if they have successfully fought to achieve them. This is the way, the proponents say, in which factory workers, women, and racial minorities have managed to secure their political rights in many countries, and this is the way in which rights ought to be achieved.

There are several critiques to be voiced against the view. First, the rights of factory workers, women and various racial groups have not been recognised in all parts of the world. Should we infer from this that members of these groups are not entitled to life, liberty and the absence of deliberately inflicted pain if they happen to live in countries which do not acknowledge these rights? Second, in many affluent countries children and people with a mental handicap are protected by legal rights which they have not achieved for themselves. Should all these rights be abrogated? Third, the ability to claim one's due may have some relevance in the context of legal and political rights. But as the foregoing examples show, the criterion cannot be employed in ethical arguments.

Protection and compassion cannot with good conscience be restricted to those who know how to earn them.

General Utility

The final argument for speciesism is an appeal to the general utility of imprisoning and studying chimpanzees, gorillas, orang-utans and other nonhuman animals for scientific purposes. Animal experimentation is the cornerstone of modern biomedicine, and the foundation of many technological advances in health care. Its discontinuation would indirectly lead to immeasurable human suffering, as the progress of medical science would come to a halt. Consequently, our present way of life would be substantially altered.

This crude utilitarian view can be challenged both factually and theoretically. One empirical point which can be queried is the actual need for animal experimentation in modern medicine and health care. Many alternative methods of examining living organisms have been developed during the last decades, and there are scientists who claim that tests involving living animals can be entirely replaced by employing these new methods. It is possible, of course, that experiments on chimpanzees, gorillas and orang-utans are irreplaceable. But it is equally possible that they are only regarded as irreplaceable by scientists who do not have adequate information concerning the alternative test methods. If this is the case, then little or no human suffering would follow from the abandonment of experiments on nonhuman great apes.

The theoretical difficulty with the crude utilitarian view is its implicit reliance on the sanctity of status quo in social life. It is, no doubt, true that the lives of many humans would be altered by the recognition and enforcement of the Declaration on Great Apes. But this in itself does not amount to an argument against the Declaration. The lives of many slave-holders were unquestionably changed by the abolition of slavery, but this does not undermine the essential validity of the new rule. Unjust and immoral social arrangements ought to be criticised even if it means inconvenience for those who have previously profited at the expense of others.

The Validity of the Declaration

It seems, then, that no outstanding and indisputably valid alternatives can be found for the mental criteria of moral equality. The second premise of our argument (P2) is, therefore, valid in the sense that it cannot be superseded by other popular views on the boundaries of equality. This, in its turn, implies that the argument as a whole is conceptually sound: assuming that the psychological facts stated in the third premise (P3) are true, both the premises and the inferences of the argument are sound. Given that chimpanzees, gorillas and orang-utans have mental capacities and emotional lives which roughly equal our own, we should not hesitate to grant equal rights to life, liberty and the absence of torture to all great apes regardless of race, gender or species.

Acknowledgement – Our thanks are due to Mark Shackleton, Lecturer in English, University of Helsinki, for revising the language of the paper.

Notes

1. See, for example, P. Singer, 'All animals are equal', in P. Singer (ed.), *Applied Ethics* (Oxford University Press, Oxford, 1986), pp. 215–28.
2. Cf. Singer, 'All animals are equal', pp. 217 ff.
3. Those who believe that all life is sacred (people like Albert Schweitzer) could argue that if trees are kicked hard enough, they suffer. These theorists may claim that we can witness the suffering of trees by observing the emergency operations by which they try to repair the damage caused by external forces. Presumably, what the proponents of the view have in mind is that we feel sympathy towards the attempts of living beings while the movements of inanimate objects leave us untouched. But this is not, strictly speaking, true. In fact, inanimate objects can also 'behave' in ways which can be interpreted anthropomorphically. When stones are thrown up from the ground, they immediately try to return to their 'natural place'. When raindrops are captured, they try to find other ways to reach the ground. And in computer games the player can often cause 'damage' which the computer tries to prevent, minimise and repair. Theoretically, nothing prevents people from feeling sympathy towards these efforts.
4. The creation of animals (two different stories) is described in Genesis 1: 26–8 and Genesis 2: 18–22.

18

A Basis for (Interspecies) Equality

INGMAR PERSSON

Ingmar Persson teaches philosophy at Lund University in Sweden. His contribution to this volume begins by asking: on what basis can we justify equality among humans? Once he has an answer to this question, he then asks whether this basis is limited to our own species, or might it also justify equality among, for example, all the great apes.

I would like to think that – to a considerable extent due to the works of some of the contributors to this book – few educated people would be inclined to dispute that things can be good or bad for nonhuman animals. At least with respect to the species with which we are here primarily concerned, chimpanzees, gorillas and orang-utans, such a denial would be especially bizarre. For the close physiological and behavioural parallels between human beings and these nonhumans constitute overwhelming evidence for ascribing consciousness and desires or interests to the latter, and it is plausible to claim that, in general, what satisfies a subject's desire is good for it, while what frustrates it is bad for it.

Of course, there is the problem of how far down the phylogenetic ladder it is correct to attribute consciousness and conative states to animals. For my own part, I am strongly tempted to attribute them not only to apes, but (at least) to all vertebrates. But this is an issue on which I need not here take a stand, it being sufficient for present purposes that the point be conceded with respect to the great apes.

There is also the difficulty that there are desires – chiefly those based on erroneous beliefs – the fulfilment of which may not be of value to the subject. But again this is a difficulty that need not concern us here, since it is hardly controversial that the satisfaction of the more basic desires which humans might share with nonhumans – desires for the elementary comforts of life, food, drink, absence of pain, etc. – is of value to the subject.

So, I take it to be perfectly safe to proceed on the assumption that life can be good or bad for at least some nonhuman beings, such as chimpanzees, gorillas and orang-utans. There are, however, many who would agree with this assumption and grant that we have some moral obligation to see to it that the lives of these beings are for the good rather than for the bad, but would balk at the suggestion that justice demands that these nonhumans lead lives that are of *equal* value to those of humans. In other words, they would deny that justice requires any form of interspecies equality. None the less, this is the proposal I shall now attempt to vindicate.

A well-known obstacle to attempts to establish that human beings are in some sense equals is the range of natural inequalities so conspicuous among them: people differ in respect of virtually every capacity or talent, mental and physical. In view of these striking differences and variations, is it reasonable to claim that all humans are equal? On the other hand, if one succeeds in overcoming this obstacle and arrives at a principle of equality according to which humans are equal, perhaps it will be found that the manifest differences between humans and non-human animals, such as the other great apes, do not preclude a wider applicability of this principle. Thus, the considerations that justify equality among humans might also justify interspecies equality between *Homo sapiens* and other species. Such is, in outline, the argument of this chapter.

The Equal Consideration of Interests

In his treatment of the topic in *Practical Ethics*, Peter Singer argues that, because of the differences alluded to, a principle of human equality cannot be based on any empirical facts about humans. 'Equality is a basic ethical principle, not an assertion of fact', he declares.[1] This basic ethical principle is 'the principle of equal consideration of interests' which lays down that 'we give equal weight in our moral deliberations to the like interests of all those affected by our actions'.[2] As Singer explains, this principle derives from his utilitarianism which states, roughly, that we should act so as to maximise the fulfilment of the interests or preferences of all beings affected. Clearly, if we are to attain this utilitarian goal, we must fulfil interests or preferences in accordance with their strength, giving priority to the stronger interests and treating equally strong preferences equally. Since not only humans, but also nonhuman animals have interests, Singer contends, this principle puts the latter within the domain of equality.

However, as Singer himself points out, there are situations in which distribution in compliance with his principle of equality will 'widen rather than narrow the gap between two people at different levels of welfare'.[3] Singer gives a case where we face two injured beings, but have sufficient medical supplies to take care of only one of them.[4] One being, A, has already lost a leg and is about to lose a toe on the remaining leg, while the other, B, is in danger of losing a leg. On the reasonable assumption that it is worse to lose a leg than a toe, the principle of equal consideration of interests recommends that in this situation we use the supplies on B, thereby increasing the difference in health status between A and B, rather than equalising them in this respect by using the resources on A.

This decision may not be ethically objectionable, but we should reflect that in *every* situation in which one can aid only one of two beings, A and B, and the increment of A's satisfaction is ever so slightly greater than that of B would be, then the principle of equal consideration of interests exhorts us to help A, *irrespective* of how much better off A might be to start with than is B. It strikes me as going against the grain to think that a principle of equality should harbour such an implication.

To be sure, this type of situation may rarely obtain in actual fact, for the more a being has, the more fulfilled it is, the harder it is to increase its fulfilment. For instance, a quantity of food that will be received with indifference by the well-fed may significantly boost the well-being of the starving. This mechanism is commonly referred to as the principle of declining marginal utility. Due to its operation, distribution according to Singer's principle of equality will as a rule bestow more on those worse off because it will generate more satisfaction there – and so it will tend to narrow rather than widen gaps in levels of welfare. This is precisely what one intuitively thinks a principle of equal distribution should do. Still, the principle of equal consideration of interests may be criticised for not guaranteeing this outcome on its own.

Moreover, Singer would not deny that the natural inequalities in the human species mean that some humans are more valuable than others as 'means' to the utilitarian end. Some possess talents and gifts that enable them to make a greater positive contribution to the good overall, by making scientific discoveries, inventions, works of art, acts of charity, etc. It may be that some of these people need to be encouraged by rewards if they are to use their talents to the full. In other words, it may be that one can justify giving these people special favours which enhance their quality of life by appealing to the fact that such treatment promotes the maximisation of satisfaction all around. On the other hand, there appear to be some who are inclined to make a negative contribution to the utilitarian goal by perpetrating crimes. From a utilitarian point of view, one can justify discouraging these by threatening them with punishments that would lower their quality of life. All in all, the natural inequalities of human beings make their value to others differ, and this seemingly provides a utilitarian justification for treating them differently, giving some better lives than others.

In general, Singer seems inclined to downplay the importance of external rewards as an incentive, declaring his sympathy with the Marxist slogan that distribution should be on the basis of need, and not at all on the basis of ability or achievement.[5] However, when he comes to discussing economy, he has to concede that so long as there is no 'decline in acquisitive and self-centred desires' of humans,[6] we shall probably have to allow private enterprise, even though it is a system which benefits the business-minded. I conjecture that had he discussed the institution of punishment, he would have had to grant that, on pain of jeopardising the maximisation of utility, it cannot be abolished, though of course it makes the lot of criminals worse.

None of this questions a rationale behind the principle of equal consideration of interests, that the prospect of fulfilling desires or interests is the sole basis of distribution of goods; it queries whether a distribution which is aimed at maximising the total quantity of fulfilment would perforce coincide with what we intuitively would regard as equality. In summary, the considerations put forward have been two. Some are healthier and stronger and will tend to live lives that *to themselves* are of greater value, and Singer's principle of equality does not always tell us to compensate those worse off (it does so only when assisted by the principle of declining marginal utility). People are also equipped with traits that make their value *to others* vary, and this leads to the principle of equal consideration of interests justifying unequal treatment of them in order not to produce a drop in the level of future fulfilment.

However, could not the last reasoning be broadened into an argument that attempts to undercut the very backbone of the principle of equal consideration of interests: namely, that satisfaction of interests is the only legitimate basis of distribution? Suppose it is argued as follows: those who make a greater contribution to the utilitarian goal *deserve* better treatment, that is, rewards, by virtue of making this contribution, while those who detract from the common good – e.g. by perpetrating crimes – *deserve* worse treatment, namely punishment. Now, the notion of desert is linked to that of *justice*: what an agent deserves is something that is of such value to the agent itself that it is *just* to provide the agent with it because it is in some sense proportionate to the value to others of what the agent has accomplished. So, in view of their different assets, justice demands that people be treated unequally.

If this argument cannot be met, it would appear that there is no hope of rectifying the results of Singer's utilitarian principle of equality, if these are found to be too inegalitarian for one's taste. For surely, if the remedy is to come from anywhere, it must come from considerations of justice. Moreover, as long as this argument stands intact, the prospects for laying a ground for interspecies equality would appear to be gloomy, since it could reasonably be argued that, by virtue of their superior assets and achievements, many humans deserve better treatment than all nonhumans.

Equality as a Principle of Justice

Fortunately, there is a reply to the last argument, and Singer himself produces the germ of it when he objects to the ideal of equal opportunity by pointing out that it rewards those who have been lucky in having inherited dispositions to behave in ways that are socially useful and penalises those who have had the bad luck of not having received such genes.[7] This brings out the salient point that no one deserves to be treated better or worse than any other because, ultimately, all our contributions to the state of the world spring from factors beyond our control and responsibility.

One cannot reasonably be thought to deserve *moral* credit or blame in proportion to the value of an event or state for which one of the following two conditions is true:

1 through action or inaction, one made no causal contribution to it *or*
2 if one made such a contribution, one either (a) could not have foreseen one's making this contribution or (b) could not have avoided making this contribution (even if one had had the requisite foresight).

One is typically ascribed moral desert for what one intentionally brings about, for this is something to which one has made a causal contribution which is foreseen and (barring rare cases of overdetermination) avoidable. The reason for these presuppositions of desert-ascription is the connection we have found between desert and justice. Clearly, if a subject makes no contribution to a state of affairs which is good/bad to others, justice cannot require that it be rewarded/punished in proportion to the value to others of this state of affairs, since it does not at all flow from the subject. Analogously, if a subject had contributed to this state of affairs, but had done so unavoidably or without being able to foresee the outcome, justice could not demand that it prosper/suffer in proportion to the value of this contribution. For then the source of this state of affairs lies partially in conditions that the subject did not handle or manipulate, but which, luckily or unluckily, made an independent impact.

However, *every* intentional action is ultimately the result of conditions to which the agent could not have made any causal contribution whatsoever. An intentional action arises out of a certain motivational state – certain desires, decisions or intentions – and certain capacities or skills. To some extent, these may have been shaped by earlier intentional actions of the agent, but – on the realistic proviso that we are not dealing with an intentional agent who has existed forever as such – we will eventually reach properties of the agent to which it could not have made *any* causal contribution, let alone one of which neither (a) nor (b) in condition 2 is true, say, properties that are determined either by early environmental influences or genetic factors. It follows that the basis of attributions of moral desert has been undermined: justice cannot require that we treat some better because their contribution to the world is beneficial and some worse because their contribution is harmful, when in the end these contributions turn out not to be theirs.

In this argument, I have presupposed determinism, but it changes little if we suppose that intentional actions are occasioned by some condition – say, a decision – that lacks a sufficient cause. For, to the extent that this decision is causally determined, it is ultimately due to causes to which one has not contributed, and to the extent that it is undetermined, it is, definitionally, out of reach of all (causal) contributing. Also, I have talked about *moral* desert specifically, but I believe the argument can easily be generalised to other types of desert, for example the beauty by virtue of which a woman is said to deserve to win a beauty contest or the high score on an IQ-test by virtue of which somebody is held to deserve a well-paid job. One's beauty or intelligence is a state of affairs to which one's contribution is slight, if not non-existent: therefore, it cannot be just to reward on the ground of it.

So, regardless of whether the world is completely determined or partially undetermined, the concept of desert lacks application: nobody deserves anything. But then what justice demands seems to be that benefits and burdens be so distributed that all end up leading lives that to them are as like in respect of quantity of value as possible. For if the different contributions people make to the good of other beings could not render it just to put some contributors in better positions than others, what could? Perhaps some thinkers are tempted to hold that it could be just that some individuals are better off than others because, according to some institutions or conventions – for example, the current institution of property – they are *entitled to* larger resources. However, this would be just only if the institutions themselves are just, and the latter appears to throw us back on the notion of desert: for instance, the relevant institution of property seems to be just only if everyone *deserves* the fruits of their labour and has a right to dispose of them as they see fit.

Since I can discern no ground for a just discrimination that is independent of desert, I conclude that, when the bases for desert-attributions have been removed, our sense of justice dictates that things be so distributed that the outcome of distribution be that all come to live lives the value of which to them are as equal as possible. The argument, then, is this:

(a) It is just that everyone (for whom there can be value) be so treated that their lives be as equal in respect of value to them as possible, *unless* they deserve to lead lives differing in value.

(b) Nobody deserves to lead a life differing in value from that of any other (since the concept of desert is inapplicable).

(c) Therefore, it is just that everyone be so treated that their lives be as equal in respect of value to them as possible.

Of course, this reasoning is not airtight since I have not *proved* – nor can I see how it possibly could be proved – that (a) is true: that is, that in default of desert, no other circumstances could make it just to let some lead better lives than others. (Note that I am not claiming that nothing could make this unequal situation the *morally better* one, because, as will soon transpire, I do not want to rule out the possibility of there being some other ethical principle that might occasionally override that of justice.) Still, as long as nobody brings such a circumstance to light, the argument stands and provides a rationale for a principle of equality or equal treatment formulated in (c). Thus, although Singer believes that 'a more through-going egalitarian principle' than his principle of equality would be 'difficult to justify',[8] some remarks of his might be read as unwittingly supplying the foundation for such a principle. Naturally, if he were to follow me in my elaboration of these remarks of his, he would have to countenance an ethical principle which has a source independent of utilitarianism.

The new principle says, it should be emphasised, how things should be from the point of view of *justice*. But considerations of justice, in this sense, could surely not be the only ethical considerations, and the new principle of equality could not be the only ethical principle. The utilitarian principle that one should maximise fulfilment has been mentioned, and to generate an acceptable morality, we most certainly need some principle of that kind – some principle which introduces considerations of *benevolence* and exhorts one to make the lives of individual beings better rather than worse. For were the principle of equality to stand alone, it would be morally indifferent whether we equalise the value of lives by raising the value of some or lowering the value of others. But, needless to say, the first strategy is morally preferable, and appeal to considerations of benevolence is needed to yield this result. (Because of these absurd consequences of a principle of pure equality, some who profess themselves to be egalitarians advocate the view that we should give *priority to the worse off*, that the increased well-being of the worse off has a comparatively greater moral weight. Even if distinct in theoretical content, the practical implications of this view seem indistinguishable from ones which can be derived from a mixture of what I have called considerations of justice and benevolence.)

Now the actual world may be such that a strict compliance with the principle of justice will bring one into conflict with this supplementary principle of beneficence or utility. One way of reasoning to this effect has been gone through above: if we do not reward those who add to the good overall, and punish those who detract from it, it is likely that the net balance of positive value will decrease in the future and that there will be less of it to go around. We might dream of a world in which everyone makes equally beneficial contributions to goodness overall, and so can be equally rewarded, but we do not live in such a world. Perhaps we can create this world – if we develop and apply (controversial) techniques of environmental and genetic manipulation. But until such a world has been attained, we must to all appearances put up with the unequal treatment of beings whom justice requires should be treated equally, unless we allow a marked drop in the production of fulfilment overall. Thus, the natural inequalities have popped up once again, but now it should be evident that they do not threaten to undermine equality of treatment as an ideal (of justice). However, they do make our ethical deliberation more complex by forcing us to weigh considerations of benevolence or utility, on the one hand, and those of justice, on the other, against each other.

Interspecies Equality

So far I have talked about (intraspecies) equality between humans, but it should be plain that the type of considerations adduced could be used to vindicate interspecies equality as well. An individual's belonging to a certain species is obviously not a result of his or her own doings; it is – given the customary criterion of species membership – something genetically fixed. Hence, a 'speciesism' that proposes to treat, for example, chimpanzees, gorillas and orang-utans worse than humans simply because they are chimpanzees, gorillas and orang-utans would be unjust.

But probably, when all is said and done, this is not what speciesism would come down to – just as racism and sexism do not simply amount to the doctrine that certain beings be discriminated against just because of their race or gender. A more intelligible speciesism (racism, sexism) proposes that beings belonging to some species (race, sex) be favoured at the expense of beings belonging to other species (races, sexes) due to characteristics *typical* of these species (races, sexes): for instance, that humans should be better catered for than all nonhuman animals because they alone are rational, have the capacity to speak a language, etc. In other words, the real basis for discrimination is not species-membership, but the possession of rationality or some other mental faculty.

Against this sort of speciesism the so-called argument from marginal cases has been marshalled: it is pointed out that if it is the absence of rationality, linguistic ability, etc., that justifies discrimination against nonhuman animals, discrimination against some humans – in particular, those who are severely mentally handicapped – is also justifiable, since they, too, lack the precious qualities. Apparently, normal chimpanzees, gorillas and orang-utans are at least as intelligent as some mentally impaired humans. It does not help the human speciesist that these humans belong to a species that *normally* is equipped with the mental assets in question, because it is surely more reasonable to treat a being according to the properties it *in fact* possesses than according to those that make up the norm for some group to which it belongs, regardless of whether or not the individual in question has them.

For the champion of animal welfare, the drawback of the argument from marginal cases is that it is powerless against theorists who are prepared to tolerate discrimination against intellectually disabled humans. However, this way out is rendered even more unpalatable by the considerations advanced in the foregoing section: it certainly cannot be just to let the mentally impaired suffer because of their disability, for this is something beyond their responsibility and control. But, plainly, the same considerations tell against treating nonhumans badly on account of their more modest mental endowments: they are of course as little responsible for their shortcomings. The proper conclusion to draw is that justice lays down that both groups be so treated that the value of their lives to them becomes as equal as possible to the value to others of their lives.

Obviously, the claim that justice demands that humans and non-humans be so treated that the value of their lives (to them) be as equal as possible does not imply that, for example, apes (or intellectually disabled humans) should be treated in precisely the same way as normal humans – for instance, offering apes the same opportunities of higher education would not boost the value of their lives by making them more fulfilled, since their mental faculties are not suited to this kind of training. It has not here been proposed that a just treatment consists in treating beings equally in the sense that they should receive the same quantities of the same goods, but in the sense that goods be so distributed that the outcome approximates as nearly as possible to a state in which all beings enjoy lives that are equally valuable or satisfying to them. And given the different endowments of different beings, this will mean supplying them with different amounts of different commodities.

Remember also that this principle is not the sole ethical principle; as has been remarked, we must also pay heed to some principle of benevolence or utility. This must be borne in mind, lest patently absurd consequences might be thought to follow from the position here advocated. Suppose that, due to the intelligence or expected life span of apes not being on a par with that of humans, their lives in their natural habitat to do not contain as much value or satisfaction as that of average humans. Then it does not follow that, all things considered, we are morally obliged to do all we can to raise the value of the life of an average ape to the level of that of a normal human being. For this policy might detract unacceptably from the *total* quantity of value achievable. (Notice, however, that the same argument tells against the proposal that we should do *everything* within our power to improve the quality of life of handicapped humans, if this should also detract unacceptably from the total quantity of value achievable.)

Moreover, considering the vast number of nonhuman species of animals which may possess consciousness and desires, some of these animals differing enormously from humans, trying to establish equality is of course an impracticable policy. I have heard it suggested that the notion of justice is not applicable to creatures with very simple minds, say, reptiles and fish, that it makes no sense to claim that it is unjust, for example, that one fish leads a better life than another. The reason offered for this is that the experiences of these beings at different times in their lives are insufficiently unified due to the absence of any articulated memory in these beings. If correct, this proposal would hope to keep the implications of equality more manageable.

But I am not convinced of its correctness: so far as I can tell, the fact that it can seem plausible may be a mere symptom of how deeply ingrained our speciesism is. If one is not guilty of the confusion of thinking that treating somebody unjustly presupposes that this being has the capacity of recognising the injustice, I do not see why beings cannot be victims of injustice unless they are equipped with the power of holding together their stream of consciousness by looking back (and perhaps forth). However, I can afford to leave this issue open, since the animals of primary interest here, chimpanzees, gorillas and orang-utans, are mentally on a level with some humans whom we emphatically take to be within the scope of ideas of equality.

The general point to bear in mind when dealing with the 'impracticability' type of objection to the principle of equality propounded here is this: in contrast to Singer's principle, it has a grounding independent of utilitarianism, therefore it must be tempered by considerations that are utilitarian in spirit to generate a sound morality. On the other hand, such a morality cannot consist solely of a utilitarian principle if the argument set out earlier is correct (namely, the argument to the effect that a distribution which widens the gap between the better and the worse off can be morally objectionable even though it maximises the total quantity of satisfaction). If there is an objection to that sort of distribution, it would seem to have to do with it issuing in a state of unjust inequality; thus, a satisfactory moral system must include a principle of equality independent of utilitarianism.

The main ethical conclusion of this chapter is that this objection could still hold good when the worse off are nonhumans, such as chimpanzees, gorillas and orang-utans – although our strong speciesist prejudices may efface this fact. So, in order to have the hope of convincing people other than those already converted, I have employed the strategy of approaching interspecies equality via intraspecies human equality, of first establishing my principle of equality for the human case and then extending it across species barriers.

Acknowledgements – I would like to thank Derek Parfit and the editors for valuable comments on earlier drafts of this paper.

Notes

1. P. Singer, *Practical Ethics* (Cambridge University Press, Cambridge, 1979), p. 18.
2. Ibid., p. 19.
3. Ibid., p. 23.
4. Ibid., p. 22.
5. Ibid., p. 36.
6. Ibid., p. 39.
7. Ibid., p. 35.
8. Ibid., p. 23.

19

Ill-gotten Gains

TOM REGAN

Tom Regan is professor of philosophy at North Carolina State University, and president of the Culture and Animals Foundation. His best-known book, The Case for Animal Rights, *presents a detailed philosophical argument for recognising the basic rights of all beings who are 'subjects of a life' – that is, capable of having individual experiences and of having their lives go well or badly. In the essay that follows, Regan considers, and rejects, various possible grounds for drawing a sharp moral boundary between human beings and the other great apes. An earlier version of this essay appeared under the same title in* Health Care Ethics, *edited by Donald Van De Veer and Tom Regan, and published by Temple University Press, Philadelphia, in 1987; it was reprinted in 1991 in Tom Regan's* The Thee Generation: Reflections on the Coming Revolution, *also published by Temple University Press. It is printed here, in a revised and shortened form, with the permission of Temple University Press.*

Late in 1981, a reporter for a large metropolitan newspaper (we'll call her Karen to protect her interest in remaining anonymous) gained access to some previously classified government files. Using the Freedom of Information Act, Karen was investigating the federal government's funding of research into the short- and long-term effects of exposure to radioactive waste. It was with understandable surprise that, included in these files, she discovered the records of a series of experiments involving the induction and treatment of coronary thrombosis (heart attack). Conducted over a period of fifteen years by a renowned heart specialist (we'll call him Dr Ventricle) and financed with federal funds, the experiments in all likelihood would have remained unknown to anyone outside Dr Ventricle's sphere of power and influence had not Karen chanced upon them.

Karen's surprise soon gave way to shock and disbelief. In case after case she read how Ventricle and his associates took otherwise healthy individuals, with no previous record of heart disease, and intentionally caused their heart to fail. The methods used to occasion the 'attack' were a veritable shopping list of experimental techniques, from massive doses of stimulants (adrenaline was a favourite) to electrical damage of the coronary artery, which, in its weakened state, yielded the desired thrombosis. Members of Ventricle's team then set to work testing the efficacy of various drugs developed in the hope that they would help the heart withstand a second 'attack'. Dosages varied, and there were the usual control groups. Administering certain drugs to 'patients' proved more efficacious in some cases than did administering no medication or smaller amounts of the same drugs in other cases. The research came to an abrupt end in the autumn of 1981, but not because the project was judged unpromising or because someone raised a hue and cry about the ethics involved. Like so much else in the world at that time, Ventricle's project was a casualty of austere economic times. There simply wasn't enough federal money available to renew the grant application.

One would have to forsake all the instincts of a reporter to let the story end there. Karen persevered and, under false pretences, secured an interview with Ventricle. When she revealed that she had gained access to the file, knew in detail the largely fruitless research conducted over fifteen years, and was incensed about his work, Ventricle was dumbfounded. But not because Karen had unearthed the file. And not even because it was filed where it was (a 'clerical error', he assured her). What surprised Ventricle was that anyone would think there was a serious ethical question to be raised about what he had done. Karen's notes of their conversation include the following:

Ventricle: But I don't understand what you're getting at. Surely you know that heart disease is the leading cause of death. How can there be any ethical question about developing drugs which *literally* promise to be lifesaving?
Karen: Some people might agree that the goal – to save life – is a good, a noble end, and still question the means used to achieve it. Your 'patients', after all, had no previous history of heart disease. *They* were healthy before you got your hands on them.
Ventricle: But medical progress simply isn't possible if we wait for people to get sick and then see what works. There are too many variables, too much beyond our control and comprehension, if we try to do our medical research in a clinical setting. The history of medicine shows how hopeless that approach is.

Karen: And I read, too, that upon completion of the experiment, assuming that the 'patient' didn't die in the process – it says that those who survived were 'sacrificed'. You mean killed?

Ventricle: Yes, that's right. But always painlessly, always painlessly. And the body went immediately to the lab, where further tests were done. Nothing was wasted.

Karen: And it didn't bother you – I mean, you didn't ever ask yourself whether what you were doing was wrong? I mean . . .

Ventricle: [interrupting]: My dear young lady, you make it seem as if I'm some kind of moral monster. I work for the benefit of humanity, and I have achieved some small success, I hope you will agree. Those who raise cries of wrongdoing about what I've done are well intentioned but misguided. After all, I use animals in my research – chimpanzees, to be more precise – not human beings.

The Point

The story about Karen and Dr Ventricle is just that – a story, a small piece of fiction. There is no real Dr Ventricle, no real Karen, and so on. But there *is* widespread use of animals in scientific research, including research like our imaginary Dr Ventricle's. So the story, while its details are imaginary – while it is, let it be clear, a literary device, not a factual account – is a story with a point. Most people reading it would be morally outraged if there actually were a Dr Ventricle who did coronary research of the sort described on otherwise healthy human beings. Considerably fewer would raise a morally quizzical eyebrow when informed of such research done on nonhuman animals, chimpanzees, or whatever. The story has a point, or so I hope, because, catching us off guard, it brings this difference home to us, gives it life in our experience, and, in doing so, reveals something about ourselves, something about our own constellation of values. If we think what Ventricle did would be wrong if done to human beings but all right if done to chimpanzees, then we must believe that there are different moral standards that apply to how we may treat the two – human beings and chimpanzees. But to acknowledge this difference, if acknowledge it we do, is only the beginning, not the end, of our moral thinking. We can meet the challenge to think well from the moral point of view only if we are able to cite a *morally relevant difference* between humans and chimpanzees, one that illuminates in a clear, coherent, and rationally defensible way why it would be wrong to use humans, but not chimpanzees, in research like Dr Ventricle's.

The 'Right' Species

An obvious difference is that chimpanzees and humans belong to different species. A difference certainly; but a morally relevant one? Suppose, for the sake of argument, that a difference in species membership *is* a morally relevant difference. If it is, and if *A* and *B* belong to two different species, then it is quite possible that killing or otherwise harming *A* is wrong, while doing the same things to *B* is not.

Let us test this idea by imagining that Steven Spielberg's E.T. and some of E.T.'s friends show up on Earth. Whatever else we may want to say of them, we do not want to say that they are members of our species, the species *Homo sapiens*. Now, if a difference in species is a morally relevant difference, we should be willing to say that it is *not* wrong to kill or otherwise harm E.T. and the other members of his biological species in sport hunting, for example, even though it *is* wrong to do this to members of our species for this reason. But no double standards are allowed. If *their* belonging to a different species makes it all right for us to kill or harm them, then *our* belonging to a different species from the one to which they belong will cancel the wrongness of their killing or harming us. 'Sorry, chum,' E.T.'s compatriots say, before taking aim at us or prior to inducing *our* heart attacks, 'but you just don't belong to the right species.' As for us, we cannot lodge a whine or a moral objection if species membership, besides being a biological difference, is a morally relevant one. Before we give our assent to this idea, therefore, we ought to consider whether, were we to come face to face with another powerful species of extraterrestrials, we would think it reasonable to try to move them by the force of moral argument and persuasion. If we do, we will reject the view that species differences, like other biological differences (e.g. race or sex), constitute a morally relevant difference of the kind we seek. But we will also need to remind ourselves that no double standards are allowed: though chimpanzees and humans do differ in terms of the species to which each belongs, that difference by itself is not a morally relevant one. Ventricle could not, that is, defend his use of chimpanzees rather than humans in his research on the grounds that these animals belong to a different species from our own.

The Soul

Many people evidently believe that theological differences separate humans from other animals. God, they say, has given us immortal souls. Our earthly life is not our only life. Beyond the grave there is eternal life – for some, heaven, for others, hell. Other animals, alas, have no soul, in this view, and therefore have no life after death either. That, it might be claimed, is the morally relevant difference between them and us, and that is why, so it might be inferred, it would be wrong to use humans in Ventricle's research but not wrong to use chimpanzees.

Only three points will be urged against this position here. First, the theology just sketched (*very* crudely) is not the only one competing for our informed assent, and some of the others (most notably, religions from the East and those of many Native American peoples) do ascribe soul and an afterlife to animals. So before one could reasonably use this alleged theological difference between humans and other animals as a morally relevant difference, one would have to defend one's theological views against theological competitors. To explore these matters is well beyond the limited reach of this chapter. It is enough for our purposes to be mindful that there is much to explore.

Second, even assuming that humans have souls, while other animals lack them, there is no obvious logical connection between these 'facts' and the judgement that it would be wrong to do some things to humans that it would not be wrong to do to chimpanzees. Having (or not having) a soul obviously makes a difference concerning the chances that one's soul will live on. If chimpanzees lack souls, their chances are nil. But why does that make it quite all right to use them *in this life* in Ventricle's research? And why does our having a soul, assuming we do, make it wrong *in this life* to use us? Many more questions are avoided than addressed by those who rely on a supposed 'theological difference' between humans and other animals as their basis for judging how each may be treated.

But third, and finally, to make a particular theology the yardstick of what is permissible and, indeed, supported by public funds in twentieth-century Western pluralistic society is itself morally objectionable, offending, minimally, the sound moral, not to mention legal, principle that church and state be kept separate. Even if it had been shown to be true, which it has not, that humans have souls and other animals do not, that should not be used as a weapon for making public policy. We will not, in short, find the morally relevant difference we seek if we look for it within the labyrinth of alternative theologies.

The Right to Consent

'Human beings can give or withhold their informed consent; animals cannot. That's the morally relevant difference.' This argument is certainly mistaken on one count, and possibly mistaken on another. Concerning the latter point first, evidence steadily increases regarding the intellectual abilities of the great apes. Much of the public's attention has been focused on reports of studies involving the alleged linguistic abilities of these animals, when instructed in such languages as American Sign Language for the deaf (ASL). Washoe, Lana, Nim Chimpski: individual chimpanzees have attained international celebrity. How much these animals do and can understand is a controversial question. Do primates have the ability to understand and use language? If they do, would they have the capacity to give or withold their informed consent? At present, no definitive answers can be given to these questions. I think it quite likely that these animals do have the necessary capacities. But it is also possible that they do not. It is not easy to trot out a doctrinaire position in this regard.

Questions about the ability of chimpanzees to give informed consent aside, however, it should be obvious that this is not the morally relevant difference we are seeking. Suppose that, in addition to using chimpanzees, Ventricle also used some humans, but only mentally incompetent ones – those who, though they have discernable preferences, are too young or too old, too enfeebled or too confused, to give or withhold their informed consent. If the ability to give or withhold informed consent were the morally relevant difference we seek, we should be willing to say that it would not be wrong for Ventricle to do his coronary research on these humans, though it would be wrong for him to do it on competent humans – those humans, in other words, who can give or withhold their informed consent.

But though one's willingness to consent to have someone do something to oneself may be, and frequently is, a good reason to absolve the other person of moral responsibility, one's inability to give or withhold informed consent is on a totally different moral footing. When Walter Reed's colleagues gave their informed consent to take part in the yellow fever experiments, those who exposed them to the potentially fatal bite of the fever parasite carried by mosquitoes were absolved of any moral responsibility for the risks the volunteers chose to run, and those who chose to run these risks, let us agree, acted above and beyond the normal call of duty – acted, as philosophers say, supererogatorily. Because they did more than duty strictly requires, in the hope and with the intention of benefiting others, these pioneers deserve our esteem and applause.

The case of human incompetents is radically different. Since these humans (e.g. young children and people with mental handicaps) lack the requisite mental abilities to have duties in the first place, it is absurd to think of them as capable of acting supererogatorily; they cannot act 'beyond the call' of duty, when, as is true in their case, they cannot understand that 'call' to begin with. But though they cannot volunteer, in the way mentally competent humans can, they can be forced or coerced to do something against their will or contrary to their known preferences. Sometimes, no doubt, coercive intervention in their life is above moral reproach – indeed, is morally required, as when, for example, we force a young child to undergo a spinal tap to check for meningitis. But the range of cases in which we are morally permitted or obliged to use force or coercion on human incompetents in order to accomplish certain ends is not large by any means. Primarily it includes cases in which we act with the intention, and because we are motivated, *to forward the interests of that individual human being*. And that is not a licence, not a blank cheque to force or coerce human incompetents to be put at risk of serious harm so that *others* might possibly be benefited by having *their* risks established or minimised. To treat the naturally occurring heart ailment of a human incompetent *is* morally imperative, and anything we learn as a result that is beneficial to others is not evil by any means. However, to intentionally bring about the heart attack of a human incompetent, on the chance that others might benefit, is morally out of bounds. Human incompetents do not exist as 'medical resources' for the rest of us. Morally, Ventricle's research should be condemned if done on human incompetents, whatever benefits others might secure as a result. Imagine our gains to be as rich and real as you like. They would all be ill-gotten.

What is true in the case of human incompetents (those humans, once again, who, though they have known preferences, cannot give or withhold their informed consent) is true of chimpanzees (and other animals like them in the relevant respects, assuming, as we are, that chimpanzees cannot give or withhold their informed consent). Just as in the case of these humans, so also in the case of these animals, we are morally permitted and sometimes required to act in ways that coercively put them at risk of serious harm, against their known preferences, as when, for example, they are subjected to painful exploratory surgery. But the range of cases in which we are justified in using force or coercion on them is morally circumscribed. Primarily it is to promote *their* individual interests, as we perceive what is in their interests. It is *not* to promote the collective interests of *others*, including those of human beings. Chimpanzees are not our tasters, we are not their kings. To treat them in ways that put them at risk of significant harm on the chance that we might learn something useful, something that might benefit others (including other chimpanzees!), something that just might add to our understanding of disease or its treatment or prevention – coercively to put them at risk of significant harm for any or all of these reasons is morally to be condemned.

To attempt to avoid this finding in the case of these animals, while holding on to the companion finding in the case of incompetent humans, is as rational as trying to whistle without using your mouth. It can't be done. As certain as it is that it would have been wrong for Ventricle to use human incompetents in his coronary research, it is at least as certain that it would have been wrong for him to use chimpanzees instead, despite the legality of using these animals and the illegality of using these humans. Here, surely, it is the law that must change, and afford the same protection for chimpanzees as it does now for humans.

The Value of the Individual

Philosophically, there is a way to insure that our gains will not be ill-gotten. This requires that we view individuals as having a distinctive kind of value – *inherent value*, to give it a name; others have called it by other names, including the *worth* or *dignity* of the individual. This kind of value is not the same as the positive value we attach to being happy or having various skills. An unhappy person has no less inherent value (no less worth or dignity) than a happy one. Moreover, the individual's inherent value does not depend on how useful others find him or her or how well he or she is liked. A prince and a pauper, a prostitute and a nun, those who are loved and those who are forsaken, the genius and the mentally handicapped child, the artist and the Philistine, the most generous philanthropist and the most unscrupulous used-car dealer – all have inherent value, according to the philosophy recommended here, and all have it equally.

To view the value of individuals in this way is not an empty abstraction. To the question 'What difference does it make whether we view individuals as having equal inherent value?' our response must be, 'It makes all the moral difference in the world!' Morally, we are *always* required to treat those who have inherent value in ways that display proper respect for their distinctive kind of value, and though we cannot on this occasion either articulate or defend the full range of obligations tied to this fundamental duty, we can note that we fail to show proper respect for those who have such value whenever we treat them as if they were mere receptacles of value or as if their value were dependent on, or reducible to, their possible utility relative to the interests of others. In particular, therefore, Ventricle would fail to act as duty requires – would, in other words, do what is morally wrong – if he conducted his coronary research on competent human beings, without their informed consent, on the grounds that this research just might lead to the development of drugs or surgical techniques that would benefit others. That would be to treat these human beings as mere medical resources for others, and though Ventricle might be able to do this and get away with it, and though others might benefit as a result, that would not alter the nature of the grievous wrong he would have done. To ascribe inherent value to competent human beings, then, provides us with the theoretical wherewithal to ground our moral case against using competent human beings, against their will, in research like Ventricle's.

Who Has Inherent Value?

If inherent value could nonarbitrarily be limited to competent humans, then we would have to look elsewhere to resolve the ethical issues involved in using other individuals (for example, chimpanzees) in medical research. But inherent value can only be limited to competent human beings by having recourse to one arbitrary manoeuvre or another. Once we recognise that morality simply will not tolerate double standards, then we cannot, except arbitrarily, withhold ascribing inherent value, to an equal degree, to incompetent humans and other animals such as chimpanzees. All have this value, in short, and all have it equally. All considered, this is an essential part of the most adequate total vision of morality. Morally, none of those having inherent value may be used in Ventricle-like research (research that puts them at risk of significant harm in the name of securing benefits for others, whether those benefits are realised or not). And none may be used in such research because to do so is to treat them as if their value is reducible to their possible utility relative to the interests of others.

Hurting and Harming

The prohibition against research like Ventricle's, when conducted on animals such as chimpanzees, cannot be avoided by the use of anaesthetics or other palliatives used to eliminate or reduce suffering. Other things being equal, to cause an animal to suffer is to harm that animal – is, that is, to diminish that individual animal's welfare. But these two notions – harming on the one hand and suffering on the other – differ in important ways. An individual's welfare can be diminished independently of causing him or her to suffer, as when, for example, a young woman is reduced to a 'vegetable' by painlessly administering a debilitating drug to her while she sleeps. We mince words if we deny that harm has been done to her, though she suffers not. More generally, harms, understood as reductions in an individual's welfare, can take the form of either *inflictions* (gross physical suffering is the clearest example of a harm of this type) or *deprivations* (prolonged loss of physical freedom is a clear example of a harm of this kind). Not all harms hurt, in other words, just as not all hurts harm.

Viewed against the background of these ideas, an untimely death is seen to be the ultimate harm for both humans and animals such as chimpanzees, and it is the ultimate harm for both because it is their ultimate deprivation or loss – their loss of life itself. Let the means used to kill chimpanzees be as 'humane' (a cruel word, this) as you like. That will not erase the harm that an untimely death is for these animals. True, the use of anaesthetics and other 'humane' steps lessens the wrong done to these animals, when they are 'sacrificed' in Ventricle-type research. But a lesser wrong is not a right. To do research that culminates in the 'sacrifice' of chimpanzees or that puts these and similar animals at risk of losing their life, in the hope that we might learn something that will benefit others, is morally to be condemned, however 'humane' that research may be in other respects.

The Criterion of Inherent Value

It remains to be asked, before concluding, what underlies the possession of inherent value. Some are tempted by the idea that life itself is inherently valuable. This view would authorise attributing inherent value to chimpanzees, for example, and so might find favour with some people who oppose using these animals as means to our ends. But this view would also authorise attributing inherent value to anything and everything that is alive, including, for example, crabgrass, lice, bacteria and cancer cells. It is exceedingly unclear, to put the point as mildly as possible, either that we have a duty to treat these things with respect or that any clear sense can be given to the idea that we do.

More plausible by far is the view that those individuals who have inherent value are *the subjects of a life* – are, that is, the experiencing subjects of a life that fares well or ill for them over time, those who have *an individual experiential welfare*, logically independent of their utility relative to the interests or welfare of others. Competent humans are subjects of a life in this sense. But so, too, are those incompetent humans who have concerned us. Indeed, so too are many other animals: cats and dogs, hogs and sheep, dolphins and wolves, horses and cattle – and, most obviously, chimpanzees and the other nonhuman great apes. Where one draws the line between those animals who are, and those who are not, subjects of a life is certain to be controversial. Still, there is abundant reason to believe that the members of mammalian species of animals do have a psychophysical identity over time, do have an experiential life, do have an individual welfare. Common sense is on the side of viewing these animals in this way, and ordinary language is not strained in talking of them as individuals who have an experiential welfare. The behaviour of these animals, moreover, is consistent with regarding them as subjects of a life, and evolutionary theory implies that there are many species of animals whose members are, like the members of the species *Homo sapiens*, experiencing subjects of a life of their own, with an individual welfare. On these grounds, then, we have very strong reason to believe, even if we lack conclusive proof, that these animals meet the subject-of-a-life criterion.

If, then, those who meet this criterion have inherent value, and have it equally, chimpanzees and other animals who are subjects of a life, not just human beings, have this value *and* have neither more nor less of it than we do. Moreover, if, as has been argued, having inherent value morally bars others from treating those who have it as mere resources for others, then any and all medical research like Ventricle's, done on these animals in the name of possibly benefiting others, stands morally condemned. And it is not only cases in which the benefits for others do not materialise that are condemnable; also to be condemned are cases, if indeed there are any, in which the benefits for others are genuine. In these cases, as in others, the ends do not justify the means.

This recognition of the *moral* equality of humans, chimpanzees and other animals who are subjects of a life is not something to which calls for *legal* reform can turn a deaf ear. The very notion of legal justice, as this applies to the treatment of human beings, grows out of acceptance of the worth, dignity or, as preferred here, the inherent value of the individual. Individual human beings, that is, if they are to be treated justly by the laws and the courts, must be treated with the respect they deserve, not because of their achievements or talents or wealth, for example, but simply because of the dignity or worth they possess as the individuals they are. Because chimpanzees (and the other nonhuman great apes) have no less a claim to such dignity, our system of legal justice must change in order to treat these animals with the respect they deserve.

Conclusion

Such a conclusion is probably at odds with the judgement that most people would make about this issue. If we had good reason to assume that the truth always lies with what most people think, then we could look approvingly on Ventricle-like research done on animals like chimpanzees in the name of benefits for others. But we have no good reason to believe that the truth is to be measured plausibly by majority opinion, and what we know of the history of prejudice and bigotry speaks powerfully, if painfully, against this view. Only the cumulative force of informed, fair, rigorous argument can decide where the truth lies, or most likely lies, when we examine a controversial moral question.

Those who oppose the use of animals such as chimpanzees in research like Ventricle's and who accept the major themes advanced here, oppose it, then, not because they think that all such research is a waste of time and money, or because they think that it never leads to any benefits for others, or because they view those who do such research as, to use

Ventricle's words, 'moral monsters'. Those of us who condemn such research do so because this research is not possible except at the grave moral price of failing to show proper respect for the value of the animals who are used.

20

The Ascent of Apes –
Broadening the Moral Community

BERNARD E. ROLLIN

Bernard Rollin holds what must be a unique combination of professorships: of philosophy, and of physiology and biophysics. He teaches at Colorado State University, where he developed the world's first course in veterinary ethics and animal rights. His book Animal Rights and Human Morality *explored the moral basis for attributing rights to animals; subsequently in* The Unheeded Cry: Animal Consciousness, Animal Pain and Science *he subjected to a withering critique the scientific ideology that animals do not really have emotions or feelings and are incapable of thought. In this chapter, he draws on the themes of both these works to make the case for an expansion of the moral community to include the great apes, and he concludes by suggesting that a court of law should examine whether the great apes have wrongly been denied the fundamental rights due to all persons.*

The Emerging Ethic for Animals

Twenty-five years ago, it would have been culturally inconceivable to suggest extending the moral community so as to include animals within the scope of moral concern. For, in the course of the 150 years during which society had paid any formal attention whatsoever to limiting human behaviour with regard to other living beings, such attention was restricted to the prohibition of overt, intentional, wilful, extraordinary, malicious, unnecessary cruelty, and the vapid encouragement of 'kindness'. This minimalistic, lowest common denominator ethic was formally encapsulated in the anti-cruelty laws, which were as much designed to ferret out sadists and psychopaths who might begin with animals and, if left unchecked, graduate to venting their twisted urges upon human beings, as to protect the animals for themselves. The traditional humane or animal welfare movement was also caught up in the categories of kindness and cruelty, and, for this reason, tended (and still tends) to simplistically categorise anyone causing animal suffering as 'cruel'.

For reasons which are not altogether clear, during the past two decades society has patently begun to move beyond the overly simplistic ethic of cruelty and kindness and has begun to reach for a more adequate set of moral categories for guiding, assessing and constraining our treatment of other animals. Perhaps the key insight behind this change is the realisation that the overwhelming majority of animal suffering at human hands is not the result of cruelty, but rather grows out of 'normal' animal use and socially acceptable motives. Scientists may be motivated by benevolence, high ideals and noble goals, yet far more animal suffering is occasioned by people acting in pursuit of these motives than by the actions of overt sadists. Factory farmers may be motivated by the quest for efficiency, profit, productivity, low-cost food and other putatively acceptable goals, yet again, their activities occasion animal suffering in orders of magnitude traditionally unimaginable.

One can venture some speculations as to why the demand for such an ethic regarding the treatment of animals has emerged only today. Primarily, perhaps, it is because society has only lately focused its attention on disenfranchised groups and individuals to a degree unprecedented in human history – women, blacks, homosexuals, native populations, the aged, the Third World, the insane, the handicapped, children, and so on. Not long ago, society had little more than an anti-cruelty ethic for the treatment of such humans. Inevitably, a generalised social concern for justice and fairness, and an emphasis on obligation rather than patronising benevolence towards the powerless and oppressed, must certainly have led to a new social look at the treatment of animals. In order to fully appreciate the answers to that question, we must recall some fundamental Socratic precepts. As a teacher and philosopher of ethics, and one who was attempting to effect meaningful social–ethical change in an intellectually sophisticated democratic society, Socrates was well aware that a moralist could not *teach* ethics, i.e. could not impart ethical truths to others the way one can impart the names of state capitals to students, and expect others to accept these truths as factual. After all, as Wittgenstein once remarked, one can take an inventory of all the facts in the universe, and never find the fact that killing is wrong. Instead, Socrates believed that a moralist can and must *remind* those whom he or she is addressing of what they have within them but of which they are not conscious, and help them, as it were, give birth to their ideas, and bring them to consciousness in a coherent way. In fact, Socrates describes the moralist-teacher's role as that of a midwife.

This Socratic notion can be extrapolated well beyond its roots in Platonism. In essence, it embodies the insight that moral progress cannot develop out of nothing, but can only build upon what is already there. In other words, the most rational and efficacious way to develop moral ideas in individuals and societies is to show them that the ideas in question are implicit consequences of ideas they already accept as veridical. In other words, I can get others to accept my ideas by showing them that these are in fact their ideas, or at least are inevitable logical deductions from ideas they themselves take for granted.

In our democratic society, the consensus social ethic effects a balance between individuality and sociality, between what philosophers call deontology and teleology, more specifically between individual rights and social utility. While most social decisions and policies are made according to that which produces the greatest benefit for the greatest number, this is constrained by respect for the individual. Our ethic builds protective fences around the individual to protect the sanctity of his or her human nature, or *telos*, from being submerged by the general or majority welfare. Thus we cannot silence an unpopular speaker, or torture a terrorist to find out where he has planted a bomb, or beat a thief into revealing where he has hidden his ill-gotten gains. These protective fences around the individual are *rights*; they guard fundamental aspects of the individual even from the general good. Specifically, they protect what is plausibly thought to be essential to being a human – believing what you wish, speaking as you wish, holding on to your property and privacy, not wanting to be tortured, etc. And they are fuelled by the full force of law.

What we have so far outlined is not difficult to extract from most people in our society. In fact, calling attention to the moral principles people unconsciously accept has traditionally been a major way of effecting social change. Arguably, something of this sort occurred when (thinking) segregationists accepted integration, or when occupations such as veterinary medicine, which had traditionally barred women, began to admit them. In both of these cases, presumably, no change in moral principles is required. What is demanded is a realisation that a moral commitment to equality of opportunity, justice, fairness, and so on, which the segregationist or person who barred women from veterinary school himself had as a fundamental commitment, entails a change in practice. In other words, to put it simply, such people had readily accepted democratic moral principles as applying to all persons. What they had ignored was the fact that the class of persons was far greater than they had acknowledged, and included blacks and women.

This mind-set, sharpened and deepened from the 1950s to the present by social demands that the ethic be truly applied in practice, not just paid lip service in theory, has informed the emerging ethic for animals.

In essence, people are moving towards applying the ethical machinery we have accepted for dealing with people to the treatment of animals. This is not to suggest that people are self-consciously doing so, any more so than most people can self-consciously articulate the consensus ethic for humans which we articulated above. Rather, they will acquiesce to a description of the ethic as applied to animals when it is articulated – in other words, when one helps them remember.

One major step towards extending the ethic to animals, not difficult for the average person to take, is the realisation that there exists no good reason for withholding it: in other words, that there is no morally relevant difference between humans and animals which can rationally justify not assessing the treatment of animals by the machinery of our consensus ethic for humans.[1] Not only are there no morally relevant differences, there are significant morally relevant similarities. Most important, most people believe that animals are conscious beings, that what we do to them matters to them, that they are capable of a wide range of morally relevant experiences – pain, fear, happiness, boredom, joy, sorrow, grief; in short, the full range of feelings which figure so prominently in our moral concern for humans.[2]

Not only does ordinary common sense accept as axiomatic the existence of consciousness in animals, it also takes for granted that animals have natures (*telos*) – 'fish gotta swim, birds gotta fly', as the song goes. Thus it is again not difficult to get ordinary people to admit that the central interests of animals' natures should be protected from intrusion; even if we use animals, animals should live lives which fit their natures. It is not an accident that major battery chicken producers do not, in their advertising, show the public how they really raise chickens; rather, they run ads showing open barnyard conditions which suggest that they raise 'happy chickens'. Ordinary people – even those who are not animal advocates – are appalled by veal calves in confinement, 'wild' animals in tiny cages, primates in austere, barren, deprived environments.

It is clear that the ethic for animals I have described, albeit in partial, tentative and fragmented form, has begun to receive codification in law, which is the key role, in my view, of the concept of 'rights'.[3] Recent United States federal legislation on animals used in research, in essence, has as its mainstay the moral right of animals not to experience pain, suffering and distress in research unless control of these states vitiates the research. To be sure, this is not codified in rights language, for animals are, legally, still property and property cannot have rights. But conceptually, the law does embody the notion that freedom from pain, distress and suffering is a key component of animal nature, as it is of human nature, and that animals are entitled to have it controlled.[4] Many animal activists see this legal protection as extremely weak, which it surely is; but they forget we are only just emerging from a situation where the only constraint on animal use – which indeed did not even apply to animal research, by legal definition – was the aforementioned prohibition against overt cruelty.

Not only does the United States law begin to acknowledge freedom *from* suffering, it has components of freedom *to*, i.e. it recognises the moral requirement to allow at least some animals to actualise at least certain aspects of their natures. The law mandates exercise for dogs, and environments for nonhuman primates which 'enhance their psychological well-being'. It is true that these requirements are aimed at certain favoured animals (a point highly relevant to including great apes in the moral community, as we shall see), and at very limited rights. But again, its revolutionary nature compared with what obtained historically regarding scientists' *carte blanche* over research animals cannot be underestimated. These rights must be respected, even though doing so is costly and burdensome to researchers.[5]

These laws also stipulate the presence of morally relevant mental states in animals, and to some extent require scientists to think in moral terms about animals, thereby undercutting the notion that science is ethics-free. The erosion of scientific ideology is amusingly illustrated by the following anecdote. When the head of APHIS (Animal and Plant Health Inspection Service), the branch of USDA charged with enforcing one of the new laws, sought help from the American Psychological Association in defining 'psychological well-being of primates', he was assured that there is no such thing. 'There will be by January 1, 1987 [the point where the law takes effect], whether you help me or not', he tellingly replied. In the wake of these laws, more articles have appeared on pain, suffering, distress, and their control in animals during the past few years than in the previous 100 years, and the scientific community has been led to accept anthropomorphising attributions, hitherto anathema, as evidence of animal pain and suffering.[6]

The second example of the codification of this new rights ethic in law comes from the Swedish agricultural animal rights law passed in 1988.[7] This law is based foursquare in the moral ideas we have outlined, and even uses rights language. In essence, the law stipulates that farm animals have the right to live their lives in accordance with their *telos*, and therefore must be kept under conditions which fit their natures. Cattle were granted in perpetuity 'the right to graze', and confinement systems designed solely for efficiency and productivity at the expense of the animal must be abandoned.

In essence, what has occurred thus far is as follows: society has gone beyond the anti-cruelty ethic, and has expressed concern that animals used by humans not suffer at our hands, and indeed, that they live happy lives. The rights of animals, as determined by their natures, must constrain and check animal treatment – convenience, utility, efficiency, productivity and expense are not sufficient grounds for overriding animals' rights. One can see this tentatively encoded in some legislation. One can also see it affecting animal husbandry without being legislated; the extensive efforts over the past decade to create zoos which respect animal natures give testimony to the spread of the new ethic. Further, it appears that society is actually willing to give up certain animal uses and conveniences for the sake of the animals: the abandonment of the Canadian seal hunt, the massive social rejection of furs, the rejection of cosmetic testing on animals by many companies, all without legislation, attest to the growing hold of the new ethic. Laws are currently in the offing in the USA to abolish confinement veal-raising and cosmetic testing, which provides further evidence of the power of the new ethic. So too does the admission by agencies charged with wildlife management that the nature of their function is changing dramatically from their traditional role of managing 'game' for hunters to serving 'non-consumptive users' of wildlife.

Solidifying and Extending the New Ethic – The Great Apes

For whatever reason, then, society has begun to 'remember' the extension of our consensus ethic for humans to animals. Like virtually all social revolutions in stable democracies, this has occurred by articulating the implicit, in an incremental fashion, rather than by imposition of radically new ideas totally discontinuous with our social–ethical assumptions. The next key question is this: 'How can one ensure that this revolution continues to unfold, rather than becoming stagnant or aborted at its current stage?'

We have already learned from Socrates something significant about ethical change. Let us now develop a number of insights from Hume. Hume pointed out that morality involves a collaborative effort of reason and passion, i.e. emotion.[8] Reason may allow us to deduce logical consequences from our moral ideas, but reason does not motivate us to act. We are motivated by emotional predilections and disinclinations, by things that make us happy, or indignant, or excite in us pity, and so on. Second, Hume pointed out that what fuels moral life is sympathy, i.e. the fellow feeling with other beings, which allows us to respond to positive and negative feelings in them as motivations for our actions.[9]

These insights of Hume seem to be borne out well with regard to our emerging ethic on animals. Obviously, the first stirrings of concern for animals were for those animals with whom we enjoy a relationship of sympathy or fellow feeling – companion animals. They respond to our moods and feelings, we respond to theirs. This, of course, helps explain the overwhelming primacy of concern of the traditional humane movement for pets, especially dogs. The further removed from us an animal is, the less likely we are to share sympathy with it. Thus many leaders of the traditional humane movement are unabashed anglers, and otherwise sensitive people feel no compunctions about dispatching snakes in any number of ways.

Furthermore, our emotions with respect to an animal, or to its treatment, will inexorably shape our tendency to apply the emerging ethic to that animal. Few of us will readily and naturally extend our ethic to sharks or rats, though patently both of these animals meet all of the criteria for moral concern: they are conscious and have natures. But we are acculturated to see these animals as noxious, as threats, as vermin, and thus we are not exercised about their wanton destruction, even when it is done simply for fun, as in the case of sharks, or in painful ways, as in the case of rodents. It is extremely significant that the general public in California opposed the hunting of mountain lions until a film of lions taking prey was disseminated, at which point concern for the lions dropped dramatically.

Emotion and empathy plainly entered into the practical extension of our ethic described above. The singling out of dogs and primates as regards to keeping them happy, not just pain-free, which occurs in the US laboratory animal law discussed above, plainly bespeaks our empathic identification with these animals, and also evidences our largely favourable emotional responses towards them. The demise of the Canadian harp seal hunt and the inevitable demise of the confined veal industry are almost certainly a function of the remarkable empathic response elicited by these paradigmatically infantile animals contrasted with the loathing that clubbing and crating elicit. It is also evident that one of the most powerful vectors resulting in the passage of the 1985 US laboratory animal laws was the release of the horrific University of Pennsylvania baboon head-injury tapes, which elicited strong emotions both on behalf of the animals and against the remote, unaffected, Nazi-like scientists, who seemed impervious to the natural revulsion occasioned in normal people by what they were doing. (The cultural stereotype of the scientist as cold and distant figure, removed from normal human feeling, which has always struggled uneasily in the public mind with the opposite stereotype of scientist as heroic fighter against disease and suffering, was doubtless also relevant here, as public disenchantment with science and technology increased during the last thirty or so years.)

This is not, of course, to suggest that the public (who is, after all, *us*), is incapable of going beyond such knee-jerk primitive emotions, but the point is that we must work to get beyond our conditioned responses; we must become educated, and be made to think. People who work with rats, for example, or virtually any animal, can develop significant respect and even affection for the animals – unfortunately, scientific ideology does its best to forestall such disruptive attachments.

In the face of these Humean considerations, it is manifest that the great apes, chimpanzees, gorillas and orang-utans, are probably the most plausible animals through which to nurture, articulate, express and solidify the emerging ethic we have described. This is true for a variety of significant reasons.

One feature of the great apes which makes them a natural locus for the emerging ethic is the extraordinary degree of fascination and, far more important, empathy, which they inspire in humans – the public response to the work of Goodall and Fossey alone attests to this. This empathy can be found in unlikely places. Some years ago, a small group seminar was held at a meeting of the American Association for the Advancement of Laboratory Animal Science, a group of researchers, animal technicians and laboratory animal veterinarians. All of these people make their living out of animal research. This particular seminar was devoted to stress experienced by people in the field. In the seminar, people told stories from their own experience illustrating points they wanted to express. The most extraordinary story was told by the chief technician at a major government research facility. Apparently, he had been closely involved in raising a baby chimpanzee who was to be used for research. Eventually, the animal was moved to a different laboratory, to be used for invasive studies. Having developed a rapport with the chimpanzee, the technician deliberately avoided following the animal's fate. One day, he was in a different portion of the institution, walking down a corridor with another technician, when the other technician called his attention to a cage, where a chimpanzee appeared to be gesticulating to them. He approached the cage, read the card affixed to it, and realised that this was the animal he had raised. It had now been used for a terminally invasive experiment. As he stood there, the animal reached through the cage, met his eyes, grasped his hands tightly, and holding on to him, died. As extraordinary as the story was, the reaction of these laboratory animal people was equally surprising, for every single member of the seminar was crying.

By the same token, most researchers who have worked with these apes develop a similar rapport. 'I won't work with them any more', one researcher told me. 'It's too painful. I can deal with monkeys OK, but these apes are too much like us – I keep seeing them as people in ape clothing.' Such a statement of course falls far short of recognising these animals as moral entities in themselves, but also clearly illustrates the unique effect that they have on triggering empathy.

I have myself experienced this response in a small way, but one which I shall never forget. One of my friends is a veterinarian at a zoo, and he invited me to tour the facility. He especially wanted me to meet the female orang-utan. It was very hot that day, and I had removed my jacket and rolled up my sleeves. As I entered the orang-utan's cage, she seized my hand in a powerful grip. Holding my left wrist, she traced her finger along a deep and dramatic scar which dominates my left forearm, while looking directly into my eyes. She then transferred her grip to my right wrist and traced the same finger along my unmarred forearm, looking at me quizzically. Then she repeated the same action along the scar. The sense that she was asking me about the scar, as a child might, was irresistible; so irresistible, in fact, that I found myself talking to her as I would to a foreigner with a limited grasp of English: 'Old scar,' I said. 'Surgery. The doctors did it.' Then I felt a wave of frustration at being unable to answer her. I confess to spending the next few hours in something of a stupor, so overwhelmed by the fact that I had, albeit momentarily, leapt the species barrier. I still cannot think or talk about that moment without feeling a chill of awe and sublimity.

Though splendidly different, these animals are like us – enough like us to trigger the essential and deep empathy so important to including them in the moral community. This natural effect has been enhanced and deepened by the work which has been done on communication with the great apes by the Rumbaughs, Premack, Patterson, Fouts and others. Leaving aside the objections of those scientists who seem hell-bent on proving that no animal can really have language, the sort of communication that does go on certainly counts as language in the minds of ordinary people. It is clear that apes can insult, joke, lie, ask, entreat, express affection and numerous emotions, grieve, teach one another, care about pets, rhyme, and so on. Such a level of communicative ability, be it language in the Chomskyan sense or not, so dramatically gives a 'window into the minds' of other animals, that it cannot fail to further augment our fundamental empathy. When this is coupled with the exhaustive fieldwork done by people such as Jane Goodall, it illustrates countless cases that can only be understood in terms of mental states like ours. As Goodall puts it:

> All those who have worked long and closely with chimpanzees have no hesitation in asserting that chimpanzees have emotions similar to those which in ourselves we label pleasure, joy, sorrow, boredom and so on . . . Some of the emotional states of the chimpanzee are so obviously similar to ours that even an inexperienced observer can interpret the behavior.[10]

Closely related to this latter point is the individuality manifested by the great apes. (As I have pointed out elsewhere, one can find significant evidence of individuality among all animals; but in the case of great apes, as in the case of humans, it cannot be missed.[11]) Whereas scientists, for example, can treat all laboratory mice as indistinguishable and interchangeable, one simply cannot do so with apes. They dramatically manifest differences in personality, temperament, preferences, and behaviour which are inescapable. Thus they tend to manifest themselves as persons, worthy of designation by proper names. Recognition of a being's individuality is a powerful spur to according that being moral concern; conversely, depersonalisation is a major step towards disenfranchisement. It is no accident that the Nazis worked very diligently to make all concentration camp inmates look alike, so that there seemed to be an endless supply of them, and individuals didn't matter.

Not only do the apes manifest emotion, personality and individuality, they manifest *reason* and *intelligence* in a manner powerfully conducive to our according them moral concern. It may well be, as philosophers such as Bentham, Singer and myself have argued, that strictly speaking rationality and intelligence are irrelevant to inclusion within the scope of moral concern as a moral object, though they are of course relevant to whether a being is viewed as a moral agent. None the less, the historical equation of intellect with worthiness of moral attention has a profound influence on our cultural mind, and every demonstration that an animal acts without thought or 'instinctively', as when a cat mindlessly scratches a tile floor after defecating, feeds the Cartesian bogeyman waiting to leap out and yell, 'See, they are just machines after all!' In the case of apes, however, as Köhler so powerfully showed more than fifty years ago in his classic, *The Mentality of Apes*, well before the advent of language studies, we have undeniable evidence of intelligence.[12] For those who erroneously equate all mentation with intelligence, the data amassed by Köhler and countless others after him in the twentieth century definitively show that at least these animals possess and demonstrate intellectual power in ways we can readily understand, and at least show them to be the intellectual equals of young, normal children. (Savage-Rumbaugh has recently demonstrated, in an article scheduled to appear in *Child Development*, that a ten-year-old chimpanzee scored significantly higher than a two-year-old child on a test of comprehension of English.[13])

These, then, are some of the basic reasons why the great apes are plausible candidates for actualising as fully as possible the emerging ethic. There are few animals so suited, both in rational and emotional terms, for fostering the widespread agreement essential to granting them 'human' rights in the context of our ethico-legal system. The question which remains, then, is how this can most expeditiously be accomplished.

One of the most significant steps must be the education of the general public regarding the extraordinary *telos* of the great apes – not so much in terms of their ability to, as it were, do humanly inspired 'tricks', such as learning sign language, as marvellous and seductive as this may be. What must be of interest is not their life in relation to us, but in itself, as something to be studied – and recounted – in its own setting, with us as restrained guests who are minimally intrusive. In this, Jane Goodall is an inspiration, displaying extraordinary courage in a world where courage is a vanishing virtue. Goodall's physical courage, in living for thirty years with chimpanzees; her scientific courage, in recounting anecdotes about chimpanzee life which, to the scientific community, are methodologically anathema, yet speak volumes; and, finally, her moral courage in defying the common sense of science's edict that science must be value-free, have already focused public attention on chimpanzees. And not only have she and others, such as Dian Fossey, who was martyred for selfless concern for the mountain gorillas, eloquently told the story of the great apes, they have also reminded us that all of these animals are members of endangered species. In this way, one can galvanise not only members of society whose primary concern with animals is as individual objects or moral concern, but also those who worry not about individuals but about the extinction of species. These disparate concerns are often at loggerheads; in this case they can effectively converge.

Those concerned about the extension of the moral community, then, must become educators, or at least patrons of education. Such education must begin in the elementary schools – what child can resist being spellbound by stories of ape life and society? It must exploit the educational potential of television and film, as Goodall has done. And it must integrate the facts with the moral message, articulating and applying the new ethic for animals.

Perhaps the best practical step one can take is to press for legislation to leave apes alone. We should not import them for zoos or for entertainment or for research, invasive or not. As Linden has pointed out,[14] we are simply incapable of respecting their natures and their attendant rights in captivity. We should let them be, and let words and cameras in the hands of the Jane Goodalls and other morally directed naturalist-scientists and artists tell of their inexhaustible wonders and grandeur. And let the dictum be proclaimed – know without hurting, see without manipulating, cherish in itself, not for myself.

Elswhere,[15] I stressed the close connection between law and morality. As long as animals were legally property, whose treatment was qualified only by vacuous prohibitions against deliberate 'cruelty', virtually all animal suffering at human hands could be countenanced by the social ethic. For this reason, talk of rights is of paramount importance, for rights, as we have seen, serve a legal as well as a moral function. Ultimately, the rights of animals, protecting fundamental aspects of their *telos*, must be 'writ large' in the legal system, if their systematic violation is to end. As we have indicated, this notion informs the emerging ethic for animals.

But it is well known that fundamental legislative change is excruciatingly slow, with that sluggishness being directly proportional to the revolutionary nature of the moral change underlying the law. Thus legislative conferral of rights for animals is a sisyphean challenge. So long as powerful vested interests oppose the change, it can become enmired indefinitely, unless public opinion can be galvanised on its behalf. This can, in my view, best be accomplished by directing current law towards the enfranchisement of animals. Such a task is of course a formidable one, since extant law basically reflects the traditional social ethic for animals. None the less, I believe it can be accomplished, specifically in the case of the great apes.

Most of the public is sufficiently familiar with recent work done on teaching language, or what is seen by most people as language, to the great apes. Though various scientists may insist that these animals do not possess genuine language, ordinary intuitions fall strongly in the other direction. After all, people can watch these animals on television, and see them putting signs together in new ways, expressing joy and sorrow, insulting and misleading researchers, and even coining new expressions. Since language is, philosophically speaking, morally irrelevant to being a rights-bearer anyway, what matters is not whether what the apes display is or is not language by some fairly abstruse scientific (or scientistic) criteria, but rather that most people who *think* that the possession of language *is* somehow morally relevant to being accorded rights *see* these animals as having language.

Consider, then, a chimpanzee or an orang-utan or a gorilla who has learned to communicate with humans using a system perceived by most people as indeed linguistic. The experiment is terminated, and the animal is no longer of use and is turned over to a zoo, or to a laboratory. When I first discussed this sort of case, in the late 1970s, this scenario was hypothetical.[16] By the late 1980s, however, it was all too real. Unfortunately, as Eugene Linden has so deftly documented,[17] this sort of case has occurred with heart-rending frequency. Linden has told of how these animals have communicated their sorrow and perplexity and anxiety and anger and fear and grief when they are wrenched out of a rich environment where they were being treated as 'honorary humans' – sometimes living as a child in the researchers' homes – and suddenly incarcerated in a place where they are used for invasive research, and have no one, human or ape, to communicate with, and live wretched, isolated, deprived, lives. Most tragic, perhaps, is that they cannot understand what they have done to merit what they, in their sublime innocence, must surely see as punishment.

Here, I have suggested, we can accelerate the moral and legal enfranchisement of animals, at least of these animals, by using the extant legal machinery, and letting them tell their story in the context of the judicial system. I am envisioning a plausible legal case based on the notions of denial of due process and cruel and unusual punishment. Surely one can make the reasonable case that these animals are, by all rational standards, *persons* who have been denied the fundamental civil rights and procedures due to persons. These animals possess measurable intelligence, sometimes in excess of that possessed by certain humans, they can reason and, most important, they can eloquently speak for themselves, and tell of their anguish and sorrow.

I am thus envisioning a new 'monkey trial', at least as spectacular in its appeal and implications as the Scopes trial, which tested the Tennessee law against the teaching of evolution. Such a trial would be extraordinarily salubrious in just the same sense. The Scopes trial forced a public airing of our scientific, conceptual and educational commitments, as well as a dialectical examination of the roles of science and religion in a democratic society. This trial would force an examination of our moral and attendant legal commitments, and illuminate areas too long left in the dark.

Whatever the outcome of such a trial, the animals would of necessity win. If the trial were lost, the issues would still have been powerfully and unforgettably aired, and the failure of our current law and morality to protect these innocent creatures forcefully and indelibly imprinted in the public mind. Indeed, even if the case never came to trial, the same result would be accomplished by the vast – and doubtless sympathetic – publicity which a skilful attempt orchestrated by first-rate legal, philosophical and scientific minds would undoubtedly generate. And, in the end, the new ethic we discussed earlier would be articulated and enlivened, to the benefit of all animals, and most assuredly to the benefit of the great apes, whose shameful treatment at human hands occasioned the need for the trial.

Notes

1. B. E. Rollin, *Animal Rights and Human Morality* (Prometheus Books, Buffalo, 1981), Parts I and II.
2. B. E. Rollin, *The Unheeded Cry: Animal Consciousness, Animal Pain and Science* (Oxford University Press, Oxford, 1989).
3. Rollin, *Animal Rights and Human Morality*, Parts I and II.

4. B. E. Rollin, 'Federal laws and policies governing animal research. Their history, nature, and adequacy', in J.M. Humber and R.F. Almeder (eds), *Biomedical Ethics Reviews 1990* (Humana Press, Clifton, NJ, 1991), pp. 195–229.

5. Ibid.

6. Rollin, *The Unheeded Cry.*

7. 'Swedish farm animals get a Bill of Rights', *New York Times,* 25 October 1988.

8. David Hume, *A Treatise of Human Nature* (edited by L.A. Selby-Bigge), (Oxford University Press, Oxford, 1888), Book II.

9. Ibid.

10. J. Goodall, *The Chimpanzees of Gombe* (Harvard University Press, Cambridge, MA, 1986), p. 118.

11. Rollin, *The Unheeded Cry.*

12. W. Köhler, *The Mentality of Apes* (1925; reprinted: Vintage, New York, 1959).

13. Forthcoming in *Child Development Monographs.*

14. E. Linden, *Silent Partners: The Legacy of the Ape Language Experiments* (Times Books, New York, 1986).

15. Rollin, *Animal Rights and Human Morality,* pp. 82–3.

16. Ibid., pp. 82–3.

17. Linden, *Silent Partners.*

21

Sentientism

RICHARD D. RYDER

Richard Ryder is a psychologist by profession; he is also one of the pioneers of the modern animal liberation movement. A former chairman of the Royal Society for the Prevention of Cruelty to Animals, and a past president of Britain's Liberal Democrat Animal Protection Group, he is currently parliamentary consultant to the Political Animal Lobby. Here Ryder urges an end to 'speciesism' – a term that he was the first to use, in a leaflet to which he refers in the opening paragraph below. His books include Victims of Science, Animal Revolution: Changing Attitudes Towards Speciesism *and, as editor,* Animal Welfare and the Environment.

Chimpanzees make love rather like humans do, but they do not usually run the risk of contracting syphilis. Not unless they are in a laboratory. An image that ever haunts me is the photograph reproduced in a Danish medical journal of the 1950s of a pathetic little chimpanzee dying of experimental syphilis, covered in skin lesions. I used it in my first two animal rights leaflets of 1970.[1]

Precisely because our chimpanzee cousins overlap more than 98 per cent of their genes with us they have been, and continue to be, mercilessly exploited in science. Their only protection has been their cost.

Chimpanzees share with us tool-making and tool-using capacities, the faculty for (non-verbal) language,[2] a hatred of boredom, an intelligent curiosity towards their environment, love for their children, intense fear of attack, deep friendships, a horror of dismemberment, a repertoire of emotions and even the same capacity for exploitative violence that we ourselves so often show towards them. Above all, of course, they show basically the same neural, behavioural and biochemical indicators of pain and distress.

Genetic engineering involving the production of new species of animals (sometimes containing *human* genes, as in the case of the Beltsville pigs and some cancer-prone mice) is making a nonsense of our traditional morality, based as it is upon speciesism. For centuries, and even today, the lay person has attached far too much importance to species differences, unaware that the boundaries between species are far from impermeable. Lions and tigers can interbreed and produce hybrids which are themselves fertile. Species of the Primate order (of which the human is a member) can also interbreed, although I know of no attested case, yet, of human interbreeding with any of the other apes: sexual attraction across species does not seem strong and mating could, at least in its natural form, prove highly dangerous for the physically weaker human partner!

Chimpanzees, gorillas and orang-utans, more than any other species, are intuitively recognised as our kin. Yet the implications of Darwinism – that biological kinship could entail moral kinship – are still resisted by vested interests and commercially motivated speciesism. It is interesting that in some instances, trading in chimpanzees for laboratory use has been an activity selected by people with an alleged Nazi background – speciesism, as it affects chimpanzees, appears psychologically close to racism.

Chimpanzees, gorillas, and orang-utans thrown down a challenge to our conventional morality. They force us to question our ethical foundations. What, then, are these? In my view, morality is about altruism. Many species show basic altruism – the protection of other members of the troop from attack, the grooming of others and, in particular, care for offspring and sharing food with other kin. Human beings show similar behaviour, but it is usually experienced as being motivated either by a (learned) sense of duty or by a spontaneous feeling of empathy based upon the awareness of others' sentiency and, in particular, their capacity to feel pain or distress. Sentiency (consciousness) is, in itself, the greatest mystery of the universe. For 100 years psychologists have fought shy of studying this ultimate phenomenon. Now it is again being examined and its similarities with quantum mechanics have been noted.[3] It is the empathic recognition that others consciously experience the mysteries of pain and distress, much as we do, that often appears to restrain our behaviour towards them.

This feeling of empathy also makes sense in evolutionary terms, and for two reasons: first, it leads to the protection and survival of offspring (and hence our genes), and second, it promotes social co-operation. Sometimes the first of these reasons, the promotion of the survival of our genes, has been emphasised to the point where it is argued that close kinship itself triggers the emotional basis for morality. But, surely, the strength of parental feelings which can exist for adoptive children undermines this argument. Biological kinship is not a necessary condition for protective behaviour, as anyone knows who has seen a cat nurturing a young rabbit or a bitch suckling a kitten. The parental potential is within us all but it can be triggered by, or directed on to, recipients who are of other species. Our kitten Leo will not only lick our old cat Albert, he will also lick me. Nobody taught him to do this. Cleaner fish and ox pecker birds are probably innately programmed to remove parasites from their dangerous but tolerant hosts of other species. Young chimps will play with young baboons. The species do not ignore each other; they interact.

Over the years the circle of morality has gradually expanded to embrace those outside the immediate circle of acquaintance. Foreigners, and those of other religions and races, have slowly been recognised as being similar to ourselves. This is not just an intellectual process but also an emotional one, and the spontaneous feeling of empathy with others appears to expand as familiarity increases. Western human beings are no longer acquainted only with other members of the family and tribe. Increased travel and the advent of television mean that we have become ever more familiar with humans of distant lands and with sentients of other species.

Children at once show sympathy for nonhumans. And they are right to do so. Other primates have been recognised as cousins since modern Europeans first become aware of them in the sixteenth and seventeenth centuries.[4] Indeed, nonhuman primates were, for nearly 300 years, exploited for entertainment in much the same way as were deformed humans, bearded ladies and other human curiosities. Now, in a world where we at least try to show a greater respect for our fellow humans, the time has come for a more general sentientism. What I mean by sentientism is simply that the importance of sentiency should be recognised and that sentiency, in whatever host it arises, should guide our morality. In chimpanzees, gorillas and orang-utans the existence of sentiency seems beyond doubt; indeed, we can be certain that members of these species can suffer just as we can. We are all related through pain. So let kinship and kindness be made one.

Notes

1. Richard D. Ryder, *Speciesism* (privately printed leaflet, Oxford, 1970); Richard D. Ryder and David Wood, *Speciesism* (privately printed leaflet, Oxford, 1970).
2. Peter Singer, *Animal Liberation*, 2nd edn (Jonathan Cape, London, 1990), pp. 13–14.
3. Richard D. Ryder, 'The mind–brain problem', *The Psychologist*, April 1990, pp. 159–160.
4. Keith Thomas, *Man and the Natural World* (Allen Lane, London, 1983), p. 132; Richard D. Ryder, *Animal Revolution: Changing Attitudes Towards Speciesism* (Basil Blackwell, Oxford, 1989), p. 72.

22

Great Apes and the
Human Resistance to Equality

DALE JAMIESON

Many of the essays in this book provide arguments for the inclusion of the great apes within the community of equals. Dale Jamieson's work begins by taking it for granted that this inclusion is justified, and asking: why do human beings have such difficulty in accepting this idea? Jamieson is professor of philosophy at the University of Colorado, Boulder, and a former director of the Center for Values and Social Policy at the same university. His early work was in the philosophy of language; since then he has written on ethical issues concerning our treatment of animals, on philosophical issues in understanding the behaviour of animals and on global warming. Together with Marc Bekoff he edited Interpretation and Explanation in the Study of Animal Behavior.

Questions about the nature and limits of the community of equals are controversial in both theory and practice. As I write these words, a bloody war between Serbs and Croats is continuing in the former Yugoslavia. Many fear that this is a preview of what may happen in what was once the Soviet Union. Tensions between Czechs and Slovaks are running high, and 'the troubles' continue in the northern part of Ireland. Here in New York, where I am writing this chapter, relations between Hasidic Jews and African-Americans in the Crown Heights section of Brooklyn have deteriorated to the point where a cycle of reprisal killings may have begun. Relations between blacks and Koreans are generally very bad, and all over America there are incidents of white racism against blacks and Asians.

Most people would express regret about all of these cases, and say that in the highly interconnected world in which we live different groups are going to have to learn to get along with each other. They don't have to like each other, but they must respect each other as equals. Whether Croat or Serb, black or Hispanic, all humans are members of the community of equals and have the right to live in peace and tranquillity, without threats to their lives and liberty.

The cases of interethnic struggle that I have mentioned pose practical problems of community: how can we bring it about that people will act on the basis of what they believe to be true and recognise the equality of others? At the level of theory the battle mostly has been won. Not many people would seriously argue that it is permissible to treat Serbs or Australian Aborigines badly on grounds of their race or ethnicity. But human beings are often better at theory than practice.

We have a long way to go even in theory towards recognising our equality with the other great apes. The idea that chimpanzees, gorillas and orang-utans should be recognised as members of our community of equals strikes many people as bizarre or outrageous. Yet, like the other contributors to this project, I believe that we have very good reasons for including them.

In this chapter I will not try to say specifically what the community of equals is or to what its members are entitled, since that has been covered elsewhere in this volume. Instead I simply endorse the general sentiments of the Declaration on Great Apes: the community of equals is the moral community within which certain basic moral principles govern our relations with each other; and these moral principles include the right to life and the protection of individual liberty.

My main interest in this chapter is in exploring why the moral equality of the great apes is so difficult for many humans to accept. What follows can be viewed as speculative diagnosis of the sources of human resistance to recognising our moral equality with the other great apes. My hope is that once the sources of this resistance have been exposed, they will to some extent have been disabled, and we can then move towards the difficult task of putting our moral ideas into practice. I will discuss what I take to be five sources of resistance to recognising our moral equality with the other great apes.

One source of our resistance may be this: we are unsure what recognising our equality with the other great apes would mean for our individual behaviour and our social institutions. Would they be allowed to run for political office? Would we be required to establish affirmative action programmes to compensate for millennia of injustices? To some extent this unclarity comes from the narrowness of our vision, and to some extent because there are significant questions involved that cannot be answered in advance. Humans often seem to have failures of imagination when considering radical social change. A world without slavery was unfathomable to many white southerners prior to the American Civil War. Life without apartheid is still unimaginable to many South Africans. One reason we may resist radical social change is because we cannot imagine the future, and we fear what we cannot imagine.

But having said this, it is true that it is very unclear exactly what recognising the moral equality of great apes would mean. Clearly it would end our use of chimpanzees in medical research, and our destruction of areas in which mountain gorillas live, but what other changes would it bring? We can benefit here from reflecting on the American experience of social change. Once slaves were emancipated and recognised as citizens, it remained unclear what exactly their rights and protections were. For more than a century various court decisions and legislative acts have continued to spell them out. This is an ongoing process, one that cannot entirely be envisaged in advance. If we are to change social practices that cannot be defended, then we must accept the unavoidable uncertainty that follows.

A second source of resistance may generally be connected to the sources of racism and sexism. Humans often tolerate diversity more in theory than in practice. The prevalence of interethnic violence and the abuse of women by men is surely related to brute differences between the groups in question. Yet the differences among humans seem slight compared with differences between humans and chimpanzees, gorillas or orang-utans. The idea of admitting our moral equality with such creatures seems outlandish in the face of such differences.

However, it is interesting to note that perception of difference often shifts once moral equality is recognised. Before emancipation (and still among some confirmed racists) American blacks were often perceived as more like apes or monkeys than like Caucasian humans. Once moral equality was admitted, perceptions of identity and difference began to change. Increasingly blacks came to be viewed as part of the 'human family', all of whose members are regarded as qualitatively different from 'mere animals'. Perhaps some day we will reach a stage in which the similarities among the great apes will be salient for us, and the differences among them will be dismissed as trivial and unimportant, or perhaps even enriching.

A third source of human resistance to equality for great apes is the lack of voices calling for such equality. The recognition of equality is deeply affected by empathy and sympathetic identification. It is difficult to identify or empathise with creatures who are remote, and whose plight is not directly articulated. Indeed the psychological importance of nearness is part of the reason why the plight of African humans is so often overlooked. Many Africans currently face famine, yet the industrialised world seems much more concerned with the less serious plights of its own victims of recession.

Even when the oppressed or disadvantaged have powerful and articulate champions, the victims themselves are often much more effective than their advocates. This aspect of human psychology has been repeatedly exploited by promoters of animal research whose public relations campaigns often feature children who claim to be alive and happy because of experimentation on animals. These individuals who have been victims of disease or disability are often more effective advocates for research than scientists. The problem with the other great apes, however, is that they are not in a position to communicate effectively with humans. As a result their case must be made by humans, and such appeals have limited efficacy.

A fourth source of the human resistance to equality is the recognition of the setback to human interests that would result. The broader the membership of the community of equals, the fewer the benefits that accrue to the members. This is part of the reason that there has been historical resistance to expanding the circle of moral concern. Societal elites have resisted claims of equality from the inferior classes; men have resisted such claims from women; and whites have resisted the claims put forward by blacks. The loss of unjust advantage is part of the cost of life in a morally well-ordered society, but those who stand to bear the cost typically try to evade it.

Perhaps the deepest source of human resistance is that claims of equality among the great apes involve a fundamental conflict with the inherited Middle Eastern cultural and religious world view of most Western societies. Judaism, Christianity and Islam all grant humans a special place in nature. In orthodox Christian views humans are so special that God even took the form of a human; it would be unthinkable that he would have taken the form of a chimpanzee, gorilla or orang-utan. Even unbelievers live with the legacy of these traditions. The specialness of humans in nature is part of the background of our belief and action. Yet, as James Rachels has powerfully argued (see chapter 15), this picture in which human uniqueness plays such an important role is being undermined by the emerging world view of science and philosophy. A secular picture which takes evolutionary theory seriously provides no support for human privilege. On this view, humans are seen as one species among many, rather than one species over many; in the long run humans are destined to go the way of other extinct species, and there is nothing in the scientific picture that directly supports the idea that this would be a loss. Of course there is no direct logical contradiction between the scientific world view and claims about human uniqueness: one can continue to hold both, as many people do. What the scientific world view does, however, is to remove much of the background which once gave plausibility to claims about human uniqueness. Without this background, such claims increasingly seem *ad hoc* and unsupported.

In this chapter I have tried to identify some of the sources of human resistance to acknowledging the moral equality of the great apes. Seen from a certain perspective, what is surprising is not that a distinguished group of scientists and philosophers are willing to assert such equality, but rather than such claims seem absurd to so many people. What I have suggested is that this initial impression of absurdity may be an expression of deep-seated fears and anxieties about our place in nature and our relations with those who are different. Even if this diagnosis is correct, such fears and anxieties will not instantly disappear. We have a long way to go before our emerging naturalistic world view will fully inform our relations with the rest of nature. But before our demons can be tamed they must be identified and understood. I have tried to take a first step towards such identification and understanding.

Acknowledgements – I am grateful to the editors and Richard Sorabji for their comments on earlier drafts of this chapter.

V

Apes as Persons

23

The Wahokies

HARLAN B. MILLER

Harlan Miller is executive secretary of the Society for the Study of Ethics & Animals. He teaches philosophy at Virginia Polytechnic Institute and State University (Virginia Tech) in the mountains of western Virginia, and is the co-editor (with William Williams) of a collection of essays entitled Ethics and Animals. *With the philosophical parable that follows, he renews a distinguished tradition in philosophical writing.*

The Speechless Tribe

Let us suppose that we discover, say in a mountain hollow in western Virginia, a group of humans bereft not just of speech but of language. They are of European stock, the descendants of a group of workers and their wives isolated in the hollow when a lumbering and railroad-building scheme collapsed in the 1840s. A century and a half of inbreeding in this small group (originally about 100, now 162) has had some striking effects in uniformity of appearance, unusual frequencies of some physical abnormalities, and so on.

But by far the most striking thing about the Wahokies (as they are called, from a mountain overlooking their small valley) is that they are wholly devoid of any language. They speak no language, read no language (even though they have preserved some newspapers and a Bible that belonged to their ancestors) and sing no songs. Wahokie toddlers do not babble, and preliminary results indicate that Wahokie children raised in ordinary families show little interest in language and make very little progress in learning to understand or to speak a language. (The oldest such Wahokie child is now six, and is far behind a normal three-year-old.) Some very limited success in teaching a few words to a few Wahokies has been attained at great expenditure of time and effort.

As far as can be determined the Wahokies have no abnormalities of the speech organs. Their vision and hearing are, on average, better than that of a normal population. For some reason, doubtless buried in brain chemistry or hardware, they just can't, or don't care to, master anything like a fully fledged human language. The cause is almost certainly genetic (no plausible environmental factors have been found) but it has not yet been identified. Comparative DNA research continues. No offspring of Wahokie/normal couplings are known to exist. In fact no such couplings have been acknowledged, though several have been rumoured.

Although the Wahokies are unquestionably human, their lack of language makes them quite unlike any other group of humans heretofore encountered. Without language they are effectively without culture. Anthropologists refuse to call them a 'tribe'. They are perhaps a band. They have no religion, and since they have no way of reckoning descent and relation, no incest taboos. Mothers and children seem to recognise a special relation, as do siblings, but that's about it. They are divided into about ten to twenty groups that merge, divide and exchange members. Most groups are controlled by a single dominant male.

The discovery of the Wahokies has created a number of most perplexing legal, administrative and moral problems. Are the Wahokies citizens? Do they own 'their' land? Are they subject to the criminal law? The state Attorney-General has ruled that they are citizens, and that any who wish to register to vote, and can show that they are at least eighteen years of age, must be allowed to do so. However, no Wahokie has shown any interest in voting, and no one has found a way of explaining government and representation to any of them.

The citizenship question is not a pressing one, but serious conflicts have arisen about the Wahokie children. A number of the youngest have been removed from their groups by county social workers and placed in foster homes (it is these cases that provide the best evidence about language acquisition, or rather its absence). Sporadic attempts were made, for a few months, to enforce the laws mandating compulsory school attendance. This was abandoned once it was realised that the school system was wholly unprepared to deal with the children.

A number of normal humans formed Friends of the Wahokies and this group has been successful in obtaining court injunctions prohibiting any encroachment on the lands the Wahokies have been occupying, forbidding any further removal of Wahokie children and suspending the enforcement of some parts of the criminal law (in particular those laws proscribing incest and setting the age of sexual consent).

The Friends of the Wahokies and other benevolently inclined normal humans often disagree about the best course of action. Some argue that

we should let the Wahokies be Wahokies and prevent outside interference. But most of these support the programmes of immunisation
and emergency medical care now being provided to the Wahokies.
Others argue that the only way to prevent this catastrophic impairment
from continuing is to prevent the Wahokies from reproducing. Perhaps
contraceptive injections or implants could be given in the course of
other medical treatment. The Roman Catholic bishop within whose
diocese the Wahokies live is of course strongly opposed to this.

In opposition to the Friends of the Wahokies, the Attorney-General,
and the bishop, who agree that the Wahokies' interests, whatever they
are, must be protected, stands the coalition headed by Jim's River
Laboratories. The coalition argues that it should be allowed to capture
the Wahokies either by removing them or by enclosing their hollows,
and to utilise them for research and testing.

The argument is straightforward. The Wahokies are humans, that is,
they are members of the species *Homo sapiens*, but they are not persons
since they are devoid of language and thus of reflective self-consciousness. They do not have the concept of a right and thus cannot have
rights. Of course they are sentient, intelligent and self-conscious in a
nonreflective way, but so are monkeys and rats. We do not hesitate to
experiment on rats in search of benefits for human persons. Nor should
we hesitate to experiment on human nonpersons for the benefit of
human persons. Since the Wahokies are in fact members of the same
biological species they are much more valuable for research and toxico-
logical testing than rats, or rhesus monkeys, or even chimpanzees. We
must not let uninformed sentiment hold back the search for truth and
safety.

Are the Wahokies Persons?

If we assume a sharp division between persons and nonpersons, and also
understand 'person' in a very strong sense, then probably the Wahokies
are not persons. If to be a person one must be capable of formulating a
life plan, of entering into fairly abstract contractual relations with
others, and of having second-order preferences about one's preferences,
then the Wahokies are not persons. And if the moral universe is divided
exclusively into the all-important group of persons and the unimportant
group of everything else, it follows that the Wahokies' interests are
vastly less significant than ours.

But the universe, physical and moral, is not like that. The interests of
persons matter and beer cans have no interests. But there are many other
sorts of things that matter. Some of these things, works of art, for
example, may have no interests of their own and yet matter morally

because persons take interests in them. More important in the present connection are the many entities that have interests of their own even though they fall short of full personhood.

A shrimp or a worm has, it appears, rather few interests, perhaps only that minimal interest in avoiding pain and seeking pleasure shared by all sentient beings. A chicken's interests are more extensive, a dog's or a cat's still more. Intelligent social beings such as wolves, monkeys and porpoises have very extensive physical, behavioural, social and (yes) intellectual interests.

The Wahokies are very intelligent, inquisitive, highly social, and sensitive beings. Let it be agreed that, due to their impairment, they fail to attain full personhood. It certainly does not follow that they are on an equal moral footing with beer cans. They are entitled to very substantial moral standing in their own right. They are, in fact, at least quasipersons.

Persons and Quasipersons

'Quasiperson' is a neologism, of course, but it refers to a sort of moral and legal status that our conceptual system has recognised, in various ways, for millennia. Some classes of human beings have been accorded a standing different from, and generally lower than, that of fully fledged persons, but higher than that of any other animals or any inanimate objects.

Slaves were generally considered quasipersons. To the extent that a society is sexist (at least in the modes with which we are familiar) women are treated as quasipersons. Infants and small children are quasipersons everywhere, as, generally, are the severely mentally impaired of any age. A quasiperson lacks the full range of rights accorded to a person, but enjoys at least most of the protections of personhood. Sometimes quasipersonhood may carry special protections not accorded full persons. Consider child labour laws, or the exclusion, under American law, of women from combat. Such protections are typically, as in these two examples, paternalistic.

In an advanced sexist, racist, society such as the Athens of 400 BC or the Virginia of 1840 with which the Wahokies' ancestors lost touch, the layers of status may be numerous. Virginia's laws then distinguished clearly and variously between adult white males (the only fully fledged citizens, i.e. legal persons), adult white females, white children, free blacks, slaves and the mentally impaired.

Multiplicity of status is not a thing of the past. Women are still subject to restrictions, some legal and many social, that do not apply to men, even in the most egalitarian modern societies. The quasipersonal status of the young and the impaired is even clearer. The very young, and those declared incompetent, are restricted and protected in a variety of ways. In general the (normal) young pass through a range of stages on the way to full personhood, some of them clearly defined by legal requirements for driving, voting, drinking, compulsory school attendance, and so on, others much more vaguely determined by custom and parental judgement.

The Wahokies are in some ways like children, in some ways like the broadly mentally impaired, in some ways like neither. Their interests, their desires and their capabilities should determine their status. Since they are incapable of understanding such notions as contract or representation it is justifiable to deny them contractual and political rights and to assign guardians to supervise their interests in such arenas. Since, on the other hand, they are quite capable of making plans, appreciating cause and effect, and expressing their preferences and aversions, their liberties should generally be restricted as little as possible. The details of their status, including the difficulties of balancing the conflicting interests of individual Wahokies, will have to be worked out politically.

But surely Jim's River Laboratories' argument will have not the slightest weight. The Wahokies are obviously sensitive, intelligent and self-conscious. Their lack of language and all that follows therefrom does not justify experimenting on them without their consent any more than it would justify such experiments on normal children of eighteen months.

Wahokies and Other Great Apes

The point of this thought experiment (or fairy-tale, if you wish) is, of course, not the moral status of Wahokies, for there are none. The point is the status of chimpanzees, gorillas and orang-utans. Chimpanzees, gorillas and orang-utans are similar in all morally relevant characteristics to the imaginary Wahokies. So, since the Wahokies are entitled to a protected quasipersonal status, these other great apes are entitled to such status as well.

One way to resist this conclusion is to reject the analogy between nonhuman great apes and Wahokies as defective in some crucial way. Are chimpanzees, say, unlike the Wahokies in a morally important respect?

It might be said that the Wahokies are our kin, genetically related to us, and chimpanzees are not. As it stands this is just not so. The Wahokies are more closely related to us than are the chimpanzees, but both are our relatives. Or rather, the Wahokies are more closely related to some of us. I am an American of European ancestry for at least ten generations, and probably more. The Wahokies are more closely related to me than to a Japanese woman with ten or twenty generations of Japanese ancestors.

But unless we are racists of the crudest sort all this is just irrelevant. That some American stranger is much more closely my kin than the Japanese woman just mentioned has not the slightest bearing on his or her moral standing. Both are entitled to my respect in their own right. Once we have escaped a narrow tribal morality we understand that moral status is determined by a being's characteristics, not its pedigree. This is as true of chimpanzees, gorillas, orang-utans and Wahokies as of normal humans.

I take it for granted that all humans have a common set of ancestors. But suppose that this were not so, that we originated from multiple parallel evolutionary paths, miraculously interfertile, or that one strand of humanity evolved and the rest were created by aliens to match. These suppositions are bizarre and wildly implausible, but, if true, would in no way impugn the moral status of any human. The question is not how we got here, but what we are like.

If, then, the Wahokie/great ape analogy is to be overthrown it must be on the grounds of some substantial difference. Wahokies look like us (some of us) and chimpanzees, gorillas and orang-utans don't (much), but it is hard to see how anyone could, with a straight face, base an argument on that. Much more plausible would be a claim that the mental life of Wahokies is of a much higher and more complex sort than that of chimpanzees, gorillas or orang-utans. Could such a claim be made out?

We cannot, of course, give comparative intelligence tests to Wahokies and chimpanzees for the simple reason that there are no Wahokies. Some might claim it to be intuitively obvious that there is much more to the intellectual difference between 'mere apes' and normal humans than the absence or presence of language. I do not find that obvious at all, and I suspect that anyone who does is underestimating the minds of chimpanzees, gorillas and orang-utans, the importance of language, or both.

There are no knockdown arguments here, only more or less well-informed predictions of what time will tell as we learn more and more about ourselves, the other great apes, and the minds and brains of all of us. No matter how much evidence accumulates, no matter how deeply a high regard for chimpanzee, gorilla and orang-utan intelligence becomes entrenched in successful science, it will be possible for speciesists to insist on an enormous gap between ape and human. It is still possible to insist on a flat earth, or on special creation of each species. In a couple of decades all these claims will be on the same footing.

[It would be possible for someone to accept the claim that the Wahokies are the moral peers of chimpanzees without renouncing the exploitation of chimpanzees in research, entertainment and so on. One need only accept the Jim's River Laboratories' argument sketched out above. The general form of my argument in this chapter is *reductio ad absurdum*. If one does not find the suggested exploitation of language-less humans morally 'absurd' (i.e. outrageously repellent) one will not be moved by this argument.]

What sorts of quasipersons are chimpanzees, gorillas and orang-utans? That is, what should their legal and moral status be? Clearly they should be protected from assault and exploitation. Killing a chimpanzee, gorilla or orang-utan should be counted as homicide just as and when killing a human is so counted. Gorillas, chimpanzees and orang-utans should be protected from harassment, physical abuse and deprivation of livelihood. They should not be experimented upon without their consent. They should not be confined or restricted except when necessary to prevent injury to themselves or others.

In practice, the best thing we can do for these apes is to leave them alone, setting aside preserves for them with strict controls on human entrance. Non-disruptive research, some degree of medical care, perhaps emergency feeding – these activities might be appropriate. The Wahokies of the story are defective humans, and it may be permissible to limit their reproduction. But chimpanzees, gorillas and orang-utans are not defective humans, they are normal chimpanzees, gorillas and orang-utans. We should wish them well, protect them from ourselves, and let them be.

24

Humans, Nonhumans and Personhood

ROBERT W. MITCHELL

Robert Mitchell teaches in the Department of Psychology at Eastern Kentucky University. He has assisted on projects studying the linguistic abilities of dolphins, a parrot, and an orang-utan, and has been actively engaged in theoretical and empirical examinations of deception, pretence, imitation and mirror self-recognition in humans and nonhumans. He co-edited (with N. S. Thompson) Deception: Perspectives on Human and Nonhuman Deceit, *the first book on deception in humans and nonhumans, and is currently co-editing (with S. T. Parker and M. Boccia)* Self-awareness in Animals and Humans, *to be published by Cambridge University Press, as well as (with N. S. Thompson and H. Lyn White Miles) a book entitled* Animals, Anecdotes and Anthropomorphism. *In this chapter he asks whether apes can be persons in the full sense of the term.*

Are chimpanzees, gorillas and orang-utans persons? In this chapter I explore this question, which is obviously relevant to the proposal that these great apes be included in the community of equals and granted some basic rights. I consider the question of personhood from a particular philosophical position in relation to nonhumans in general, and then discuss implications of my findings for the proposal.

In his novel *The Day of the Dolphin*, Robert Merle presents a view of dolphins as persons – that is, as self-conscious beings with some control over their own activities, who reflect (via language) about these activities and have a moral sense.[1] It was this image which initially prompted me to assist on projects to teach dolphins (and a parrot) language-like communication. Although after working with these animals I was at a loss as to whether they were persons, the research led me to try to find in nonhumans evidence of the concomitants of personhood: verbal communication, self-reflection and the knowledge that others are persons. This last kind of knowledge can only come about through some system of communication which allows for expression of self-consciousness.[2]

On one persuasive view of what it is to be a person, that I am a person requires, at some point in the development of personhood, that I recognise that you recognise that I have consciousness. Thus, there seems to be a triple reflection of consciousness necessary for personhood: 'The ego, the *I*, cannot truly emerge . . . without doubling itself with an *ego in the eyes of the other*.'[3]

These and other requirements for personhood[4] are neatly arranged in a conceptual scheme by Dennett.[5] In his analysis, personhood derives first from three mutually interdependent characteristics: being rational, being intentional and being perceived as rational and intentional. Once a being is acknowledged to have these three characteristics, personhood requires that the being reciprocate by perceiving others as rational and intentional; next the being must be capable of verbal communication and finally of self-consciousness. These last three characteristics are hierarchically dependent, building upon the first three.

None of these characteristics (except the last) need be recognised as such by the being, and, as Dennett suggests, most intelligent beings exhibit the first four. Thus, the big problem in discerning whether a being is a person is discerning whether the being communicates verbally and is self-conscious. With most mature human beings, verbal communication and self-consciousness seem obviously present, but with nonhumans and some humans, both characteristics are not obvious. By self-consciousness, Dennett means that one is capable of reflective self-evaluation, that is, of 'adopting toward *oneself* the stance not simply of communicator but of . . . reason-asker and persuader'.[6] Dennett bases his definition of verbal communication upon Grice's theory of non-natural meaning, which entails that, by producing some utterance, an utterer intends for another to recognise the utterer's intention for the other to do or believe something as a result of the utterance.[7] Only if there is evidence of verbal communication should we expect evidence of self-consciousness. The question thus becomes, given that nonhumans are without speech, how are we to discern any of their (potentially) verbal communications?

The best answer was, I think, provided by Bateson[8] in his analysis of metacommunication in nonhuman play.[9] Bateson was concerned with the evolution of verbal communication, and wondered how a non-linguistic being could develop a system of communication that could lead to communication of the human sort. He suggested that the being could simulate its activities, and make the fact of simulation apparent to other beings. Bateson believed, for example, that monkeys playfighting were acting *as if* fighting (that is, were simulating fighting), yet were indicating that they were not fighting by making it evident that they were not fighting. Although Bateson's analysis of playfighting by most monkeys may be inaccurate, his suggestion that recognition of one's own or another's simulation is a way in which nonlinguistic beings could develop verbal communication is intriguing.[10] Is there any evidence that nonhuman beings recognise, create and/or communicate via simulation?

Such evidence generally involves intentional imitation.[11] For example, the sign-taught orang-utan Chantek imitated a two-dimensional photograph of a gorilla pointing to her nose.[12] To perform this imitation, Chantek must have known how he would look when he performed the action depicted in the visual image, as well as how it would feel to create this action with his own body. He must have been able to translate from the visual image to his own kinesthetic sensations – that is, to sensations of his own 'bodily position, presence, or movement'.[13] This translation of a visual image to a kinesthetic act which resembles (simulates) the visual image is intriguing in that it implies that Chantek has a cross-modal representation of his body, which itself implies that Chantek has an imaginal representation of himself. Often such cross-modal imitation is in the form of pretence: a rhesus monkey carried and repositioned a coconut shell in direct imitation of a rhesus mother's carrying and repositioning her infant,[14] and the sign-taught chimpanzee Washoe bathed a doll as her human care-givers had bathed her.[15] These pretend imitations again imply that the imitator has a capacity for translation between visual experiences and kinesthetic representations of him/herself, such that the imitator presumably could know how to effect actions based upon a visual mental image of him/herself engaging in an action. The surprising thing about most instances of nonhuman imitation and pretence is that there is no aspect of communication: the animal seems content to be engaged by the simulation without any attempt to engage another in the fact of simulation. Can beings who imitate their visual experiences create visual experiences based on imitation for other beings? That is, can they communicate via simulation?

While there are few instances of cross-modal imitation in nonhumans, there are even fewer instances of communicative imitation, that is, of simulation which one being uses to inform another of its simulativeness and thereby to metacommunicate intentionally and produce non-natural meaning. Communication of non-natural meaning is, of course, a direct test of verbal communication, and one which usually involves simulation of some sort. An illustration of the complexities which can be presented via simulation is present in an example of a twelve-month-old boy J who gets his father from another room to retrieve a block that landed behind the bookcase when the block had flown off the top of the boy's jack-in-the-box. Because his father does not understand what the boy wants (not having seen the original launching), J recreates the block's trajectory behind the bookcase:

J . . . takes his father's hand in his own, places them both on top of the jack-in-the-box, makes a kind of explosive noise, and moves his and his father's hand in an arc toward the bookcase. J then reaches his own hand down behind the bookcase, making somewhat conventionalised effort sounds to signal reaching. Still looking at his father, he says something like, 'Block'.[16]

J intends his father to get something which has been shot behind the bookcase, and he also intends his father to recognise this intention, and he realises these intentions by simulating for his father the events which led to the block's current unavailability. Similar re-enactments which create communication via simulation and non-natural meaning are performed by nonhumans. An Indian Ocean bottlenose dolphin, after attempting to get the attention of human observers outside a glassed-in area of her tank, used imitation to convey a shared experience: after observing 'a cloud of cigarette smoke', the dolphin 'immediately swam off to [her] mother, returned and released a mouthful of milk which engulfed her head, giving much the same effect as had the cigarette smoke'.[17] Similarly, the orang-utan Chantek eschewed using the sign for *milk* and instead recreated part of his normal milk-getting situation: he 'gave his caregiver two objects needed to prepare his milk formula and stared at the location of the remaining ingredient'.[18] These communications via non-natural meaning required simulation of a previous event. Such communication would become extremely cumbersome if it were the only means of information transfer, and clearly non-natural meanings must give way, and did in the cases of J and Chantek, to conventionalised utterances if there is to be a continuing and extensive communication system.[19] But what is striking about these instances is that the simulator intends for the other to recognise that the simulation is *about* something which resembles, but is other than, the actions themselves. The simulator intends the observer to recognise the resemblance to something, and to recognise that the simulator intends that the observer should recognise the resemblance. The communication of non-natural meaning is part of intentional simulation recognised as such, much as Grice[20] would have it.

Why would beings resort to anything as convoluted as communication of non-natural meaning via simulation? The answer may be that animals had to communicate with two audiences at once: for one audience information was hidden, for the other information was manifested. For example, adult rhesus monkeys appear to use playfighting as a threat to other monkeys when a direct threat would be problematic:[21] they playfight in such a way as to intimate real fighting to one monkey (their 'play partner') while appearing to be merely playfighting to an ally of that monkey (e.g. the monkey's mother) so as to avoid the ally's intervention. A similar deceptive manoeuvre seems present when a simulation is used simultaneously to hide information and present misinformation in relation to the *same* individual. For example, a gorilla acted as though foraging to get near an infant whose mother was very protective;[22] a chimpanzee imitated friendly facial expressions and gestures to lure another chimp near enough to be able to attack her easily, and another chimpanzee imitated play to distract others from aggressive acts toward himself.[23] In all of these instances of deception, the being had to retain one interpretation for him/herself and present saliently another interpretation for another – *the same action came under two descriptions for the animal*. With such dual-description, the being is able 'to recognise that the other individual's and its own signals are only signals, which can be trusted, distrusted, falsified, denied, amplified, corrected, and so forth'.[24] This dual-description has significant consequences for morality, because

> *If I am to be held responsible for an action* (a bit of behavior of mine under a particular description), I must have been *aware* of that action under that description. Why? Because only if I am aware of the action can I *say* what I was about, and participate from a privileged position in the question-and-answer game of giving reasons for my actions.[25]

So with intentional deception via simulation comes the capacity for dual-description, and from communication of non-natural meaning via simulation comes a sharing of perspective. Given that some animals can satisfy criteria for verbal communication, we can now look for evidence of self-consciousness in these animals, with its attendant sense of moral responsibility.

Clearly the instances of intentional deception, imitative pretence, and communication of non-natural meaning suggest that the imitator has some sort of internal representation of self, and/or some sort of internal representation of the other's psychology, both or either of which are used to guide behaviour. But these activities do not seem to evidence the sort of self-consciousness we are concerned with, in which the being evinces reflective self-evaluation. Are there other sources of evidence which might indicate such reflective self-consciousness?

One traditional avenue for discerning self-consciousness is recognition of oneself in a mirror. Mirror self-recognition is present in many humans, chimpanzees and orang-utans, and in a few gorillas, and is commonly taken to be a sign of pre-existing self-consciousness.[26] Recognising oneself in a mirror implies recognising a simulation of one's own body, which suggests a capacity to understand simulation as such, as well as its relation to one's own body. Once achieved, mirror self-recognition entails that the being recognises that an action the being experiences kinesthetically is identical to the visual display of that action in the mirror,[27] a capacity which is already evidenced in imitative pretence. Indeed, it is likely that this ability to recognise simulation in a mirror is based, in part, upon a previous ability to imitate activities of other beings via kinesthetic–visual matching.[28]

Note, however, that the mere fact of recognising oneself in a mirror would not be indicative of self-consciousness in the sense of thoughtful self-evaluation.[29] Knowing that one looks like one's image in the mirror does not mean that one has capacities for reflecting about one's situation in life or evaluating oneself. But some responses to mirrors *do* indicate some sense of critical self-evaluation, in that the observer uses the mirrored image of self to create an image of self which is aesthetically or culturally satisfying to others or oneself. In humans, a 'reflective self-awareness' which takes into account our awareness of how others perceive us is utilised in making such an ideal image of ourselves.[30] Such reflective self-awareness seems absent in the other great apes. Reflective self-awareness differs from the self-awareness present in the examples of imitative pretence, intentional deception and communication of non-natural meaning in that it incorporates awareness of another's awareness into one's awareness of self: a final condition of personhood. The term 'reflective' is used to imply both that one's self-image is experienced through others' perspectives (i.e. the self-image is 'reflected back' from the other) and that one is capable of self-examination (i.e. one can 'reflect on' and evaluate one's thoughts).

Although true for both humans and apes that 'By means of the image in the mirror [one] becomes capable of being a spectator of himself', it may be true only for humans (and not even for all humans) that with one's self-image 'appears the possibility of an ideal image of oneself – in psychoanalytic terms, the possibility of a super-ego'.[31] Because of reflective self-awareness, the ideals of morality are possible. But along with such reflective self-awareness comes the ability to make a deliberate argument in support of one's moral vision.[32]

So far it is clear that nonhuman beings, including the great apes, are not persons, in that they lack full self-consciousness, or what I am here calling reflective self-awareness. It would appear that humans, but not apes, because of reflective self-awareness 'can ponder past and future and weigh alternative courses of action in the light of some vision of a whole life well lived'.[33] But the great apes seem to differ from human beings in this way by degree rather than in kind, in that their self-awareness and perspective-taking provide them with mental images which represent themselves and others, and they can use these images to plan their activities.[34] To plan is not merely to have a prospective image, but to imagine oneself *within* a prospective image. Thus, the simulator can imagine different scenarios by which he or she can choose to live, and in this sense has the beginnings of reflective self-awareness. Chimpanzees (and other great apes) may not be able to 'formulate a general plan of life',[35] but can formulate a general plan for (at least) a day or a night: for example, a chimp can select and carry a tool which will assist in obtaining food at a distant location, or carry clumps of hay for warmth when moving from her inside enclosure to the outside which she had experienced as cold the day before.[36] These plans for the day can include plans for their offspring, for example that the youngster should learn manual skills through imitation of a parent's demonstration.[37] Thus, great apes can ponder past and future and weigh alternative courses of action in the light of some vision of a whole day or night well lived.

In many ways, the capacities great apes show in relation to awareness of themselves, awareness of others' psychology and reflective self-awareness indicate that they (at least in our present state of knowledge) are much like young children. In the same way that we would protect children from torture, provide them with (a restrained) freedom, and guarantee their right to life, we must provide the same conditions for the great apes.[38] It is true that apes cannot make a deliberate argument for their rights,[39] but neither can young children or oppressed people whose oppressors refuse to learn their language; yet morally we protect their rights, at least in principle.

Still, there are problems. One has only to read Hearne's[40] analysis of social interaction between the chimpanzee Washoe and her human caregivers to recognise that even sign-taught chimpanzees are unlikely to become as integrated as do human children and dogs into a human-centred culture (although see Patterson and Linden's[41] presentation of apparently easy social interactions with the sign-taught gorilla Koko). If apes are to be considered persons, it is unclear whose standards are to be applied in correcting apes' behaviour, given that they do not appear to have their own moral standards and are unlikely to accept or understand fully those of any given human culture. Even if, given appropriate human counsel, apes began to develop the morality of a given human culture (as some might argue occurs when signing apes use the signs BAD and DIRTY in particular situations), it seems inappropriate to induce them to form such moral judgements. Apes, unlike children, do not require contact with human beings to develop naturally and to live their lives in accord with their daily plans. Thus, our ethic toward apes might allow them to be free to live within specified boundaries. Yet problems are likely to occur over such boundaries when apes and humans encroach upon each other's territories, much as boundary disputes occur among humans. As with conflicts between humans, conflicts between apes and humans (or among apes) can lead to murder. Given that chimpanzees, for example, hunt and eat both humans and other chimpanzees,[42] it is unclear how one is to settle disputes: should a chimpanzee be held responsible for the murder of another chimp, or of a human, whom the chimpanzee has killed for food? If so, how is such responsibility to be accounted for legally? Among humans,

> those who desire to rule over others must give justifying reasons for their rule, which allows critics . . . to analyze the reasons and expose any flaws. For chimpanzees no such rhetorical deliberation is necessary, and thus there is no ground for moral criticism [of chimpanzees].[43]

Unfortunately, any 'moral vision' or sense of 'justice' which is possible within the constraints of ape mentality is egocentric and pragmatic,[44] and does not involve argumentation and deliberate debate. The fact that criticism of the behaviour of chimpanzees and other apes on moral grounds is impossible has serious consequences, in that apes cannot be held accountable for their actions. (I am a bit disturbed here with the parallels between claims of apes' incapacities for moral action, and assumptions of learned eighteenth- and nineteenth-century white men that non-white and/or non-male persons were inferior and thus should not be given equal political power.[45] However, the differences between apes and humans in linguistic skill are, and were, clearly not found between white males and other humans, and these skills – or similar ones – seem necessary for reflective self-awareness. Still, one powerful result of the present proposal to include apes in the community of equals is to make quite salient apes' similarities to humans and especially, in our current state of knowledge, to human children.)

Clearly, the fact that great apes are not fully persons creates difficulties in our treatment of them: although it is easy and reasonable to grant the right to life and protection from torture to these apes, the right to liberty is more ethically cumbersome. Human beings murder other human beings, and can be held accountable because they have chosen to violate the liberty of another – a moral transgression. Because apes have no rules against murder, any curtailing of their liberty as a result of their murdering another – or even to prevent a potential murder of another – creates moral difficulties if apes have the status of persons without the responsibilities. We can hold a person responsible for his or her actions because he or she can recognise the (legal and moral) consequences of these actions and give reasons for the goodness of these actions. Because apes are not persons in this full sense of the term, they cannot be held accountable because they cannot understand morality and give reasons for their actions. Thus, some restrictions upon their liberty with the effect of avoiding their death or curtailing murder of them can be morally defensible because we humans value our own and their lives. (Such curtailment is also practised, of course, toward children and some intellectually disabled, immoral or amoral older human beings.) Although great apes are not persons in the full sense of the term, they have psychological capacities which make them ends-in-themselves deserving of our protection.

Acknowledgements – I appreciate the advice offered by the editors, which greatly improved the clarity of my essay.

Notes

1. R. Merle, *The Day of the Dolphin* (Simon and Schuster, New York, 1969).
2. A. C. Danto, 'Persons', in P. Edwards (ed.), *The Encyclopedia of Philosophy*, vol. 6 (Macmillan and Free Press, New York, 1967), pp. 110–14; W.R. Schwartz, 'The problem of other possible persons: dolphins, primates, and aliens', *Advances in Descriptive Psychology*, vol. 2 (1982) pp. 31–55.
3. M. Merleau-Ponty, 'The child's relations with others', in J.M. Edie (ed.), *The Primacy of Perception* (Northwestern University Press, Illinois, 1960/1982), pp. 96–155.
4. Danto, 'Persons'.
5. D.C. Dennett, 'Conditions of personhood', in *Brainstorms: Philosophical Essays on Mind and Psychology* (Bradford Books, Cambridge, MA, 1976/1978), pp. 267–85.
6. Dennett, 'Conditions of personhood', p. 284.
7. H. P. Grice, 'Meaning', *Philosophical Review*, vol. 66 (1957) pp. 377–88.
8. G. Bateson, 'A theory of play and fantasy', in *Steps to an Ecology of Mind* (Ballantine Books, New York, 1955/1972), pp. 177–93; G. Bateson, 'The message "This is play"', in B. Schaffner (ed.), *Group Processes: Transactions of the Second Conference* (Josiah Macy Jr. Foundation, Madison, NJ, 1956), pp. 145–242.
9. R. W. Mitchell, 'Bateson's concept of "metacommunication" in play', *New Ideas in Psychology*, vol. 9 (1991) pp. 73–87; H. P. Grice, 'Meaning revisited', in N. V. Smith (ed.), *Mutual Knowledge* (Academic Press, New York, 1982), pp. 223–43.
10. Mitchell, 'Bateson's concept', pp. 77–8.
11. R. W. Mitchell, 'A comparative-developmental approach to understanding imitation', in P. P. G. Bateson and P. H. Klopfer (eds), *Perspectives in Ethology*, vol. 7 (Plenum Press, New York, 1987), pp. 183–215.
12. H. L. W. Miles, 'The cognitive foundations for reference in a signing orangutan', in S. T. Parker and K. Gibson (eds), *'Language' and Intelligence in Monkeys and Apes: Comparative Developmental Perspectives* (Cambridge University Press, Cambridge, 1990), p. 535.

13. W. Morris (ed.), *The American Heritage Dictionary of the English Language* (American Heritage and Houghton Mifflin, Boston, 1969), p. 721.
14. J. A. Breuggeman, 'Parental care in a group of free-ranging rhesus monkeys (*Macaca mulatta*)', *Folia Primatologica*, vol. 20 (1973) p. 196.
15. R. A. Gardner and B. T. Gardner, 'Teaching sign language to a chimpanzee', *Science*, vol. 165 (1969) p. 666.
16. S. Rubin and D. Wolf, 'The development of maybe: the evolution of social roles into narrative roles', in E. Winner and H. Gardner (eds), *New Directions for Child Development*, No. 6: *Fact, Fiction, and Fantasy in Childhood* (Jossey-Bass, San Francisco, 1979), p. 18.
17. C. K. Tayler and G. S. Saayman, 'Imitative behaviour by Indian Ocean bottlenose dolphins (*Tursiops aduncus*) in captivity', *Behaviour*, vol. 44 (1973) p. 291.
18. Miles, 'Cognitive foundations for reference', p. 535.
19. R. G. Millikan, *Language, Thought, and Other Biological Categories: New Foundations for Realism* (Bradford Books, Cambridge, 1984).
20. Grice, 'Meaning revisited', pp. 233–4.
21. J. A. Breuggeman, 'The function of adult play in free-ranging *Macaca mulatta*', in E. O. Smith (ed.), *Social Play in Primates* (Academic Press, New York, 1978), pp. 169–91.
22. R. W. Mitchell, 'Deception in captive lowland gorillas', *Primates*, vol. 32 (1991) pp. 523–7.
23. F. de Waal, 'Deception in the natural communication of chimpanzees', in R. W. Mitchell and N. S. Thompson (eds), *Deceptions: Perspectives on Human and Nonhuman Deceit* (SUNY Press, Albany, 1986), pp. 221–44.
24. Bateson, 'A theory of play and fantasy', p. 178; see also Grice, 'Meaning revisted', pp. 233–4.
25. Dennett, 'Conditions of personhood', pp. 282–3.
26. G. G. Gallup, Jr, 'Self-awareness and the emergence of mind in primates', *American Journal of Primatology*, vol. 2 (1982) pp. 237–48; F. Patterson, 'Self-recognition in gorillas', Paper presented at symposium on *Gorilla Cognition and Behavior* (American Society of Primatologists meeting, Davis, California, 1990); but see R. W. Mitchell, 'Mental models of mirror self-recognition: two theories', *New Ideas in Psychology*, vol. 11 (1993, forthcoming).
27. Mitchell, 'Mental models'.
28. P. Guillaume, *Imitation in Children*, 2nd edn (University of Chicago Press, Chicago, 1926/1971); see also Mitchell, 'Mental models'.
29. Mitchell, 'Mental models'.
30. Ibid.
31. Merleau-Ponty, 'The child's relations with others', p. 136.
32. L. Arnhart, 'Aristotle, chimpanzees, and other political animals', *Social Science Information*, vol. 29 (1990) pp. 477–557.
33. Ibid.

34. R. W. Mitchell, 'A framework for discussing deception', in R. W. Mitchell and N. S. Thompson (eds), *Deception: perspectives on Human and Nonhuman Deceit* (SUNY Press, Albany, 1986), pp. 3–40; R. W. Mitchell, 'A theory of play', in M. Bekoff and D. Jamieson (eds), *Interpretation and Explanation in the Study of Animal Behavior*, Vol. 1: *Interpretation, Intentionality, and Communication* (Westview Press, Boulder, CO, 1990), pp. 197–227.

35. Arnhart, 'Aristotle, chimpanzees, and other political animals'.

36. J. Goodall, *The Chimpanzees of Gombe: Patterns of Behavior* (The Belknap Press of Harvard University Press, Cambridge, MA, 1986), pp. 31, 587–8.

37. C. Boesch, 'Teaching among wild chimpanzees', *Animal Behaviour*, vol. 41 (1991) pp. 530–2.

38. Mitchell, 'A framework for discussing deception', p. 30.

39. Arnhart, 'Aristotle, chimpanzees, and other political animals', p. 526.

40. V. Hearne, 'A walk with Washoe: how far can we go?' in *Adam's Task: Calling Animals by Name* (Knopf, New York, 1986), pp. 18–41.

41. F. Patterson and E. Linden, *The Education of Koko* (Holt, Rinehart and Winston, New York, 1981).

42. Goodall, *The Chimpanzees of Gombe*, pp. 282–5.

43. Arnhart, 'Aristotle, chimpanzees, and other political animals', pp. 526–7.

44. F. B. M. de Waal, 'The chimpanzee's sense of social regularity and its relation to the human sense of justice', *American Behavioral Scientist*, vol. 34 (1991) pp. 334–9.

45. See P. Singer, *Animal Liberation: A New Ethics for Our Treatment of Animals* (Avon Books, New York, 1977), pp. 1–4; S. J. Gould, *The Mismeasure of Man* (W. W. Norton, New York, 1981), pp. 32–5.

25

Personhood, Property and Legal Competence

GARY L. FRANCIONE

What is the legal status of the great apes, and how might it be changed? If the Declaration on Great Apes were to be implemented, how could the rights of apes be protected? In this chapter Gary Francione discusses these questions. His examples come from American legal cases, but his suggestions are broadly applicable. Francione is professor of law at Rutgers University School of Law, in Newark, New Jersey, and director of the Rutgers Animal Rights Law Clinic.

Legal Rights for Great Apes

The Declaration on Great Apes requires that we extend the community of equals to include all great apes: human beings, chimpanzees, gorillas and orang-utans. Specifically, the Declaration requires the recognition of certain moral principles applicable to all great apes – the right to life, the protection of individual liberty, and the prohibition of torture.

If these principles are going to have any meaning beyond being statements of aspiration, then they must be translated into legal rights that are accorded to the members of the community of equals and that can be enforced in courts of law. Indeed, the Declaration itself suggests that moral principles would be enforceable in courts of law.

Those who support these principles as *legal* rules, and not just as moral statements, must recognise that the laws of most countries, and certainly American law, present very serious conceptual obstacles to such a position. The American legal system is replete with categories and attaches negative consequences to those categories based on race, sex, sexual preference, age, nationality and disability.[1] But there are no more serious consequences than those attached to classification based on *species*.

For example, although experimentation involving a human being requires the person's informed consent (or the consent of a legal guardian if the person is incapable of giving consent), and is subject to legal scrutiny on a number of different levels, animal experiments (in the United States) may be performed on any animal for any purpose that is approved by a committee of other animal experimenters, and the concept of informed consent obviously has no applicability. Moreover, there is no need – as there is in virtually every instance of human experimentation – to demonstrate that the experiment will benefit the experimental subject. Once some being is placed on the other side of the species barrier, the law provides virtually no protection for that being, and humans may harm that being in ways that would be unthinkable if applied to even the most disadvantaged members of human society.

In this chapter, I want briefly to examine the notion of animals as property to explain the species discrimination reflected in virtually all legal systems. Next, I will argue that even the most conservative understanding of the concept of equal protection requires that all great apes be regarded as 'persons' under the law. Finally, I will examine some problems concerning the integration into the legal system of rights for all great apes.

Animals as Property

The reason for the differential treatment accorded to nonhumans has to do with the fact that as far as the legal system is concerned, animals and humans occupy completely different positions. Human beings are regarded by the law as capable of having rights; nonhumans are regarded as incapable of having rights. Although there is an increasing social awareness about nonhuman animals and a consensus that animals possess at least some moral rights that ought to be recognised by the legal system, animals still have the status of being the *property* of human beings – just as slaves were once regarded as the property of their master, or women as the property of their husbands or fathers.

There is, however, general consensus that animals ought not to be subjected to 'unnecessary' pain or 'unjustified' killing. Although animals are viewed as property that cannot possess rights, there are many laws that purport to provide some level of protection for animals in a variety of different circumstances. The problem is that when humans try to determine whether suffering or death is 'necessary', they inevitably engage in 'hybrid' reasoning in which they balance human interests, including the legal fact that humans are regarded as having rights, and especially rights in property, and animal interests, which are unsupported by accompanying claims of right.[2] And nonhumans are a form of property that humans seek to control. Under this framework, animals can virtually never prevail as long as humans are the only rightholders and animals are merely regarded as property – the object of the exercise of an important human right.

The treatment of animals as property is illustrated in a legal case, *State v. LaVasseur*.[3] Kenneth LeVasseur was an undergraduate student at the University of Hawaii. In January 1975, he began to work at the university's marine laboratory at Kewalo Basin, Honolulu, as a research assistant. His primary responsibilities involved repairing and cleaning the laboratory's dolphin tanks, and feeding and swimming with the dolphins. In May 1977, after working with the dolphins for over two years, LeVasseur decided that the dolphins were in great danger as the result of their confinement in the laboratory tanks. He and several other people removed two dolphins from their tanks at the laboratory and released the animals into the ocean. LeVasseur was charged and convicted of first-degree theft, and he appealed.

The primary issue in LeVasseur's appeal was whether the trial court had erred in ruling that LeVasseur could not use a 'choice of evils' defence. This defence, which has different formulations depending upon the jurisdiction, provided under Hawaii law that certain conduct, otherwise criminal, could be justified if the actor believes that conduct 'to be necessary to avoid imminent harm or evil to himself or to *another*' if '[t]he harm or evil sought to be avoided by such conduct is greater than that sought to be prevented by the law defining the offense charged'.[4]

LeVasseur argued that he was trying to prevent greater harm to *another* in two senses. First, he argued that the dolphins should be included within the term *another*. The appellate court rejected this argument because the statute defined 'another' as a *person*, and although corporations and associations can be considered as 'persons' under the law, the court ruled that dolphins could not be so considered.

Second, LeVasseur argued that the term 'person' was also defined under Hawaii law to include the United States. LeVasseur maintained that the policy of the federal Animal Welfare Act[5] was to prevent cruelty to animals, and that by releasing the dolphins, LeVasseur was trying to protect the humane treatment policy of the United States. Although the court accepted that the Animal Welfare Act 'and its accompanying regulations manifest a national policy to protect the well-being of laboratory animals like the instant dolphins',[6] the court held that LeVasseur had acted improperly because he should have contacted the federal government and reported the life-threatening condition of the dolphins, and should not have deliberately chosen theft as his means of helping the animals.

In the court's view, the crime of theft of property was as great an evil as the evil that LeVasseur sought to prevent – the death of the dolphins. This decision is completely understandable given the fundamental premise of the Animal Welfare Act and other current legislation concerning animals – that nonhumans are the *property* of humans, and can be exploited for human benefit. Given a characterisation of nonhumans as property, the researchers at the university were only exercising one of their *rights*, the right of private property. It should not, therefore, be surprising that even though the court recognised that the policy of the federal Animal Welfare Law was to treat animals humanely, the evil that LeVasseur had sought to avoid – the *in*humane treatment of the animals – was no greater (and, indeed, was a lesser evil) than the evil that he had actually caused – the violation of the university's *property* rights.

The animal interest, even when it is substantial from the animal's point of view, is virtually always accorded less weight than the most trivial of human interests because, for the most part, human beings have absolutely no way of looking at nonhumans except as some form of property. Most human/animal conflicts arise because some human is trying to exercise his or her rights of property over some nonhuman, and the conflict ostensibly requires that we balance the human and animal interests. In doing so, however, we are comparing the interests of humans, which are supported by claims of legal right, and especially the legal right to exercise control over property, with the interests of nonhumans, which are unsupported by claims of legal right *because* the animal is regarded as the property of the human whose interest is at stake.

This balancing of completely dissimilar but peculiarly related legal entities accounts for why animals virtually always lose in the balance. For example, we condemn the 'unnecessary' suffering of animals, but we tolerate the use (which is synonymous with 'the abuse') of chimpanzees in circuses. There is no way that the use of chimpanzees in circuses can be squared with our rejection of 'unnecessary' animal suffering without understanding that such animal abuse is made 'necessary' merely by the existence of the right of property in the chimpanzee – and in Western societies, that property right is seen as a very powerful right.

Legal Personhood

If the Declaration on Great Apes is to have any meaning as far as chimpanzees, gorillas and orang-utans are concerned, then it is necessary that the concept of *legal personhood* be extended to them, and they must cease to be treated or viewed as the *property* of humans. It is only then that apes may be regarded as legitimate holders of *legal* rights.

Some may argue that the concept of legal personhood cannot, as a conceptual matter, be extended to anything but human persons. Indeed, it is the common lay view that humans have legal personhood and that only humans can be persons. A brief examination of legal doctrine, however, demonstrates that this view is incorrect. Not all humans are (or were) regarded as persons, and not all legal persons are human.

Slaves in the United States and elsewhere were clearly human, but did not enjoy legal personhood; they were regarded as property in much the same way that nonhuman animals are regarded today.[7] Similarly, women in the United States were once regarded as the property of their husbands, and in some nations, women still suffer significant legal disabilities. Children have certain rights and are not, strictly speaking, the property of their parents; they are, nevertheless, disabled under the law from full legal personhood.

Just as not all humans are regarded as persons, not all persons are human. In the *Le Vasseur* case, the defendant argued, in part, that the definition of 'another' should include dolphins because 'another' would include corporations and the exclusion of the dolphins was unjustifiable. Under common law, corporations are regarded as 'persons' with full rights to sue, be sued, hold property, and so on. Indeed, it would not be an exaggeration at all to suggest that much American law concerns the activities of corporations, and the practice of most American lawyers contains at least some corporate work. When an economic system finds it advantageous, its notion of 'personhood' can become quite elastic.

Bioethics is currently preoccupied with the question of legal personhood as that term applies to foetuses and to the incompetent elderly. The Supreme Court has held that personhood may not, as a matter of constitutional law, be extended to a foetus that is not viable, and viability usually occurs in the third trimester of pregnancy.[8] An earlier incarnation of the legal person would violate the woman's right of privacy, which includes her right to terminate an unwanted pregnancy. Similarly, there is a great deal of litigation and discussion about when 'death' occurs in the ill or elderly for purposes of determining when life-support measures may be withdrawn by the family or the state.

What is peculiar about many of the discussions of legal personhood is that the attributes of personhood often the focus of debate as to whether this or that being is a 'person' are *clearly* present in *all* great apes. For example, one of the more exhaustive sets of attributes of human personhood is presented by bioethicist Joseph Fletcher, who sets out a list of fifteen 'positive propositions' of personhood. These attributes are: minimum intelligence, self-awareness, self-control, a sense of time, a sense of futurity, a sense of the past, the capability of relating to others, concern for others, communication, control of existence, curiosity, change and changeability, balance of rationality and feeling, idiosyncrasy and neocortical functioning.[9] Although we may doubt that chimpanzee, gorilla or orang-utan foetuses (or even very young chimpanzees, gorillas or orang-utans), or the incompetent elderly chimpanzees, gorillas or orang-utans exhibit all of these attributes, we are no longer able to doubt that *all* great apes (except foetuses, and perhaps the very young or the incompetent elderly) possess these characteristics.

Moreover, the great apes possess these characteristics in substantially similar ways. That is, there is a high degree of similarity among the great apes in terms of mental capabilities and emotional life – characteristics which, for most of us, are central to the notion of 'personhood'. And it is in this respect that exclusion of any great ape from the community of equals must be viewed as being arbitrary and irrational, and not merely morally unjustifiable.

Philosophers such as Tom Regan[10] and Peter Singer[11] have demonstrated convincingly that there can be no moral justification for what Richard Ryder has called 'speciesism', or the determination of membership in the community of equals based upon species. Many opponents of the attack on speciesism offer the supposed rebuttal that if we reject species as a criterion for determining membership in the community of equals, it will be impossible to 'draw the line' as it were. Although I personally find this position unconvincing, and believe that a coherent moral view requires that we draw the line at *sentience*, which would result in the inclusion within the community of equals of a broad range of beings, I also recognise that it is difficult to accuse someone who disagrees with my view of being 'irrational' (simply on the basis of that disagreement).

Wherever we decide to draw the line, however, it is clear that all great apes belong on the same side, and that it would be irrational to place some great apes on one side, and some on the other. Interestingly, such an approach is thoroughly consistent with the most conservative of the tests employed in the interpretation of equal protection guarantees under American law. That is, when someone challenges a government classification as violative of equal protection, the challenger has the burden of showing that the classification is irrational and not related to any legitimate government interests. (There are instances when the government has the burden of demonstrating a compelling interest, and the government's claim is subject to strict judicial scrutiny. This more stringent test is applied when the classification affects a fundamental right, such as the right of free expression. There are also other tests that fall in between the 'irrationality' test and the 'strict scrutiny' test. For example, classifications based on gender receive 'heightened' scrutiny.)

Any classification of 'person' that excludes some great apes entirely from the sphere of 'personhood' is, based on the clear and indisputable mental and emotional similarity among *all* great apes, thoroughly irrational. This irrationality is exacerbated by the fact that the United States Constitution permits 'persons' to include thoroughly dissimilar entities, such as corporations.

In this sense, the argument for including great apes within the scope of our moral concern is a most powerful one. The argument does not require the inclusion of all sentient beings within the scope of our moral concern as *persons*, but only requires that we include those beings who are so substantially similar to human beings that their exclusion would be completely irrational – as irrational as creating a classification of human beings based on hair colouring.

Guardianship and Criminal Liability

There are two further issues that I wish to raise in connection with the issue of inclusion of all great apes when viewed from a legal perspective.

(a) The Use of Guardians to Protect Legal Rights

Critics of the inclusion view may argue that if all great apes are given certain fundamental legal rights, it will be impossible to enforce these rights because only humans can resort to the use of courts to enforce their rights. The standard response to this objection is that we could appoint guardians for those great apes who are unable to assert their own legal claims. For example, many people with profound mental disabilities and children are human beings who cannot themselves vindicate their legal rights. In such cases, courts appoint guardians for these 'wards', and the guardians represent their interests in court.

Although I generally agree with this 'standard response', there are clearly some difficulties with the guardianship model that have as yet gone unaddressed. One concern involves identifying who (i.e. what human) will be assigned the task of serving as the guardian. In the cases of people with mental disabilities and children, the legal system assumes that some family member can be identified who we assume will act in the best interests of the ward. This assumption may be unwarranted in many cases, and the family member may not always have the ward's interest at heart, but, at least in theory, and in most cases in practice, the system does work.

In the case of chimpanzees, gorillas and orang-utans, on the other hand, it seems difficult to provide criteria for identifying the appropriate guardian. Under the system now in existence, nonhumans are owned, but we certainly cannot rely on the owner of the animal to act in the nonhuman's best interests; indeed, as discussed above, the human/animal conflict generally arises in the context of the assertion of a property right over an animal by the owner or someone acting on behalf of the owner. In a community of equals, nonhumans will not be owned at all, so the solution is completely inapplicable.

One possible answer is to entrust the guardianship to people or organisations who have demonstrated an interest in, or knowledge about, animal issues. For example, we might entrust guardianship to members of animal protection organisations. Such an approach would, however, engender endless controversy and dispute even among animal advocates, who often disagree about what *is* in the best interests of nonhumans.

One possible answer to this difficulty may be found by carefully describing the rights accorded to nonhuman great apes, thereby limiting the range of discretion that would need to be exercised. As the range of discretion is limited, the identity of the guardian becomes less important. That is, in the case of human 'wards', legal issues generally concern what is in the 'best interests' of the ward. These issues are often very complex because it is not always clear what is in the 'best interests' of a human. For example, if a guardian has to determine whether to place the minor ward in a different school, or the mentally disabled ward in a different institution, it may not be clear, even after much investigation, what is in the 'best interests' of the child or the mentally disabled person.

If, however, we conscientiously provide to great apes the rights articulated in the Declaration – the right to life, liberty and freedom from torture – then, in most instances, we will know what is in the 'best interests' of the nonhuman ape. Indeed, a 'strict construction' of these rights will only serve to benefit the nonhumans because the Declaration is tantamount to an assertion that certain identified states of affairs *are* in the 'best interests' of great apes. For example, if we accept that there can be *no* unwarranted interference with the liberty of any great ape, we can no longer tolerate the incarceration of these nonhumans in zoos or research laboratories. Accordingly, the only role of the guardian would be to seek the immediate release of the great ape unjustifiably restrained or imprisoned. Of course, it may be necessary to resocialise the nonhuman under certain circumstances, but there is far less disagreement about the methods of resocialisation than there is about whether these animals may be incarcerated at all.

The difficulties in determining what is in the 'best interests' of a great ape arise only when 'speciesism' creeps in, and we provide for the treatment of chimpanzees, gorillas and orang-utans in a way that might conceivably differ from the treatment of humans. This is not, of course, to say that a strict application of the three basic rights articulated in the

Declaration will make legal determinations easy. Under the best of circumstances, there will be very difficult questions, and there will probably be many of them. The point is only that these questions become easier as we understand the rights of nonhuman great apes as absolute *prohibitions* on what can be done to these beings.

(b) Criminal Liability for Great Apes

The Declaration envisages circumstances where a great ape can be deprived of liberty for committing a crime. If this reference to criminal culpability is intended to apply to humans, and not to chimpanzees, gorillas and orang-utans, then I see no difficulty with the notion. Alternatively, if apes may be detained or incarcerated if they pose a threat to the community, then that notion may also be acceptable under at least some circumstances. For example, if, for whatever reason, a gorilla presently imprisoned in a zoo cannot be returned to the wild, some form of detention may be justified for the safety of both the gorilla and the community.

It is clear, however, that the Declaration would prohibit resurrecting formal criminal liability for *any* nonhuman (great ape or rat).[12] However intelligent chimpanzees, gorillas and orang-utans are, there is no evidence that they possess the ability to commit crimes, and in this sense, they are to be treated as children or mental incompetents. Such treatment is consistent with the use of the guardianship model to facilitate the incorporation into the legal system of rights for nonhuman great apes. We have guardians who represent the interests of wards because the wards are deemed, for whatever reason, to be incapable of making responsible choices for themselves. So too, use of the guardian model for nonhuman great apes recognises that these nonhumans lack certain capacities, and one such capacity is the ability to comprehend and use legal rules. It would seem most unjust and unsound to recognise these incapacities for purposes of appointment of a guardian, and then to permit criminal liability to be imposed.

Conclusion

The Declaration of Rights is a sensible attempt to recognise what we have for too long ignored: that certain nonhumans *must* be regarded as 'persons' for purposes of obtaining legal protection of their fundamental rights. Indeed, not to accord such protection to all great apes is irrational in light of the demonstrated mental and emotional similarities among all great apes. It is, moreover, particularly unjustifiable under a legal system that already regards some nonhuman entities as legal persons. These nonhuman entities are regarded as persons not because they share any salient aspect of personhood; rather, their status is derived from the need for modern capitalistic legal systems to provide for investor protection. If, however, we regard the term 'personhood' even in a weakly objectivistic manner (i.e. as a concept with determinative conditions of application) there can be no doubt that personhood is a term that must be applied to all great apes.

Notes

1. See M. Minow, *Making All the Difference: Inclusion, Exclusion and American Law* (Cornell University Press, Ithaca, NY, 1990).
2. I am currently at work on a book in which I explore more fully the difficulties presented when animals are treated as property.
3. 613 P.2d 1328 (Haw. Ct. App. 1980).
4. 613 P.2d at 1332 (emphasis added).
5. The Animal Welfare Act is the primary American statute concerning the treatment of animals used for experiments. It is found in Title 7 of the United States Code, at sections 2131 to 2157.
6. 613 P.2d at 1333.
7. See A. Watson, *Slave Law in the Americas* (University of Georgia Press, Athens, GA, 1989).
8. *Roe v. Wade*, 410 U.S. 113 (1973).
9. J. Fletcher, 'Humanness', in *Humanhood: Essays in Biomedical Ethics* (Prometheus, New York, 1979), pp. 12–16.
10. T. Regan, *The Case for Animal Rights* (University of California Press, Berkeley and Los Angeles, 1983).
11. P. Singer, *Animal Liberation*, 2nd edn (New York Review of Books, New York, 1990).
12. See E. P. Evans, *The Criminal Prosecution and Capital Punishment of Animals* (Dutton, New York, 1906).

26

Great Apes as Anthropological Subjects – Deconstructing Anthropocentrism

BARBARA NOSKE

Barbara Noske has an MA in cultural anthropology and a doctorate in philosophy from the University of Amsterdam, where she has held a research post in the Department of Social Philosophy. Presently she is working as an independent publicist on human–animal and culture–nature issues, as well as women's studies. Noske is a rare kind of social scientist: she includes animals within her field study. In her book Humans and Other Animals *she challenged the assumption that anthropological methods are applicable only to human beings. Here she considers the possibility of a new kind of cultural anthropology: one that includes within its scope not only human beings, but also the great apes.*

Anthropologists would commonly define their discipline as the study of *anthropos* (humankind) and would therefore think it perfectly natural to pay little or no attention to the nonhuman realm of apes. Of course, apes may figure in cultural anthropological studies now and then but would do so mainly as raw material for human acts and human thought.

Numerous anthropological studies do indeed dwell upon humans or human groups in their relation to animals: animals have frequently been treated as an integral part of human economic constellations and human ecological adjustments[1] and as part of all sorts of human belief systems and world views.[2] Anthropology has a long tradition of studying the ways in which human groups and cultures deal with and conceive of their natural environment, including other species, as seen in the theories of cultural ecology and cultural materialism. Such studies usually confine themselves to humans in their capacity as agents who act upon or deal with animals or in their capacity as subjects who think and feel about animals. Consequently, animals tend to be portrayed as passive objects that are acted upon and thought and felt about. Far from being considered subjects in their own right, i.e. as beings who are exercising a more or less independent influence upon their surroundings (which may or may not include human beings), the animals themselves are virtually overlooked by anthropologists.[3] At least they tend to be considered unworthy of anthropological interest. The reason for this, as I see it, is to be found in the view commonly held by social and cultural anthropologists, namely, that animals in themselves would have nothing to offer a science which is concerned with 'the social' and 'the cultural'. (Physical anthropologists who study the biological aspects of humankind and the origins of culture are exceptions.)

Anthropologists and sociologists, as well as other scholars in the humanities generally, assume that sociality and culture do not exist at all outside the human realm. These phenomena are taken to be exclusively human, a view which lands anthropologists (as well as their colleagues in other social sciences) in the circular argument that animals, not being human, cannot possibly be social or cultural beings. Although there are many different definitions of culture,[4] most anthropologists would agree that culture is something which transcends the purely organic level, something which is generated and sustained by collectives of people in ways not directly monitored by the genetic make-up of each and every individual involved. Social and cultural anthropologists tend to characterise humans in terms of the material and social arrangements these humans make and by which they are also shaped: as beings who socially constitute and are constituted. Besides appearing as a material and observable object or event, culture also encompasses non-material and unobservable ideational codes such as symbols, concepts and values which are collectively, if not always wittingly, created by people.[5] Humans are taken to make their own history and while their natural history was once believed to be made for them, modern humanity increasingly tries to shape that history as well. Animals, by contrast, are believed to have only a natural history which is made for them and which has caused them to evolve in the first place.

Unlike human beings, animals tend to be regarded as organisms primarily governed by their individually based genetic constitutions, i.e. by their instincts or by their genes. But this conviction turns out to be a rather *a priori* one, given that almost no student of human society and culture ever pauses to ask the same questions about animals that they ask about humans. One simply does not look for the social and the cultural where surely it cannot be found, i.e. outside the human sphere! If one preconceives humans to be the sole beings capable of creating society, culture or language, one will thereby have pre-empted 'ape' forms of society, 'ape' culture and 'ape' language almost by definition. Paul Bohannan is among the very few anthropologists who do think animals worthy of anthropological consideration; he warns against the widespread bias towards a fundamental human–animal discontinuity. Bohannan feels that animals and animal qualities should be a matter of examination rather than of definition and assumption.[6]

On the whole, animals continue to figure in anthropology not only as objects for human subjects to act upon but also as antitheses of all that according to the social sciences makes humans 'human'. The social sciences tend to present themselves pre-eminently as the sciences of discontinuity between humans and animals. This is not to say that anthropologists and sociologists reject all notions of humanity as an animal species among other species. Most anthropology and sociology textbooks even start with some sort of exposé on the origins and evolutionary background of humankind. They often embark upon a brief discussion of the prehistorical ape–human passage (an overture to physical anthropology) and will at least list some human biological universals such as a number of bodily traits and life-cycle activities which humans have or have not in common with the apes.[7] Formally, at least, humanity's relatedness to other animals, particularly the apes, is being acknowledged and its morphological and physiological resemblance to great apes taken note of, albeit in rather hasty descriptions.

There are very few cultural anthropologists, however, who seem willing to ask what animal–human continuity might mean in terms of their own field. Neither do most anthropologists question the common hierarchical subject–object approach to the human–animal relationship, and least of all do they pose questions as to the ways in which animal subjects might relate to human subjects. By far the majority of anthropologists and sociologists tends to treat our continuity with the great apes as some kind of purely material residue from a remote prehistorical past. At the most our 'primate' body is taken to have provided the material base upon which our real 'humanness' (mind, sociality, culture, language) could arise.[8] To many social scientists our humanness represents an animal basis of sorts plus a vital addition.

At the same time, social scientists tend to be very much on their guard against any form of biological essentialism, that is, against notions of humanness, a universal human nature.[9] They will hasten to point out the dangers of explaining social differences between people in terms of their biological essences such as race or sex – and rightly so – and will stress that it is humanity's nature to have no specific nature. It is frequently argued that anthropology should study the social, cultural and historical diversity *within* the human species rather than come up with some sort of universal human essence.[10]

Ironically, many social scientists who hold this view almost imperceptibly gravitate towards precisely those essentialist positions they claim to detest so much, as soon as another biological category comes into view, namely the species barrier between us and the apes. Suddenly rather clear-cut notions as to what *is* human and what *is* animal crop up among anthropologists and other social scientists. Their outspoken and valid criticisms of those who think in terms of biological essences suffer from considerable credibility loss in the face of their own assumptions about human and animal essences. Implicitly, anthropologists do have conceptions pertaining to a universal human essence: it seems first and foremost to be embodied in our 'nonanimalness' and in the ape's 'nonhumanness'. But if humanness should be identical with nonanimalness, then what constitutes animalness, what are animals and what are apes? As we have noted before, hardly any social scientist shows even a remote interest in animals for their own sake, let alone cares to ask sociological and anthropological questions about animals. What I am referring to are typically questions pertaining to collective meanings as existing within animal societies in general and within great ape societies in particular. What kind of social concepts and codes do chimpanzees, gorillas and orang-utans have, what sociocultural patterns do individual apes adhere to?

Given the exclusion of animals from their respective fields, what grounds do social scientists have for making such confident statements about animals, especially about what these are not? What conceptions do these scientists have of animals and where did they get their ideas?

Animals as Outcomes of Biological Laws

Elsewhere[11] I have pointed out to what extent the present image of animals has really been shaped by natural sciences which tend to be reductionist and objectifying in their approaches. The so-called 'Scientific Revolution' created a specific image of the natural world: nature as a semi-divine life force was replaced by nature as a useful, technical object with no purpose or value of its own. In the eyes of modern scientists, nature became mechanical, measurable and quantifiable matter. This process, known as *the disenchantment of nature*, has been strengthened by the advocation of the *parsimony principle*, a principle which requires scientists to reduce natural phenomena to their lowest possible, that is, most material and measurable, level. According to this principle, scientists should try to stick to material levels of explanation, such as neurophysiological or genetic ones, instead of opting for explanations of a mental, let alone sociocultural, nature.[12] In the animal sciences this has led to a neglect of the complexity of animals and animal life, since cognitive priority is given to those aspects of animals which can be measured and controlled. The other aspects of animals, those which can hardly be measured in terms of mainstream biological methodology, are therefore dismissed as secondary or, worse, rationalised out of existence. Thus, the part of animals under scrutiny in the laboratory and under the control of the positivist natural scientist comes to represent the whole animal. On top of all this came the Cartesian notion of the animal-machine, a view which denied animals all subjectivity, feeling, suffering, needs, fear or knowledge.[13] In short, animals ended up as passive and law-bound products of the laws of living matter.

Even within the mainstream paradigm of animal science, Darwin's ninteenth-century theory of evolution and its twentieth-century extensions such as sociobiology, animals tend to be portrayed as outcomes of the blind-working mechanism of natural selection. Darwin himself still left some room for purposeful action, judgement and choice on the part of the animal as well as for tentative animal cultures as *causal* factors in evolution. But, on the whole, Darwinians consider social and cultural behaviour as a *result* rather than as a *cause* of evolution. From the perspective of modern biologists animals are primarily passive vehicles for genes. According to this view animals will inadvertently exhibit that behaviour which will promote their genes into the next generation. As a result all animal behaviour is treated as connected with adaptation and gene transmission: an individual animal behaves toward others in a way which has proved adaptive.[14]

Note that in this view all crucial animal properties are conceived of as hereditary, that is, they are taken to reside 'in the genes'. This is inevitable if one sees the natural selection of genes as the engine behind all animal life. After all, genes are present in individual organisms, not in cultures. And it is by means of the individual's reproductive organs that genes are transmitted.

Mainstream animal science such as biology and ethology is just not designed to handle things that are socially and culturally created (instead of genetically) and which in their turn shape their creators. Generally speaking, biologists and ethologists do not possess the methodological equipment to conceptualise the non-material aspects of culture such as ideas, meanings and values held by groups. Needless to say that many biologists run the risk of giving biological deterministic explanations for human *and* animal behaviour.

When, but *only* when such biological reductionisms are directed at humans, social scientists are up in arms. In contrast, these students of human society and culture seem to uncritically endorse whatever animal image is being put forward by animal scientists. What social scientists typically fail to appreciate is whether or not this animal image really reflects the 'truth' about animals. Contrary to the images of 'man' and more recently the images of 'woman', there has as yet been very little debate on the image of animals as a product of human construction.

Anthropocentrism or Anthropomorphism: A Dilemma?

Indeed, many students of animal behaviour tend to impart a very mechanistic sort of animal image. Judging from their comments (at least when they speak professionally), animals do not prefer, want or love, let alone create culture, but instead are endowed with a certain behaviour caused by some mechanism and triggered by some stimulus. It is easy to see why this objectified imagery of animalness is deterring social scientists from even wishing to confront the thorny issue of human–animal continuity. Thus, anthropologists are reluctant to face up to the possibility that not only humans but apes too might create forms of culture worthy of anthropological study. Since the biological sciences, with the compliance of the social sciences, have been passing themselves off as the expert animal sciences, animals (including apes) have become associated with purely biological and genetic explanations. Generally, anthropologists feel that it would not do to associate humans with these kinds of explanations. In order to safeguard humans from another onslaught of biological determinism anthropologists are generally quite defensive about the nonapeness of humans. Insofar as humanity's apeness is acknowledged, it is acknowledged only on a physical level. But here we are not so much concerned with anthropology's acknowledgement of humanity's 'apeness', but first and foremost with the discipline's recognition of sociality and 'culturality' among apes. Could great apes fully be subjects for anthropological research?

Underlying the failure to acknowledge ape–human continuity in its cultural capacity is yet another factor: the fear among biologists of being accused of anthropomorphism, of attributing exclusively human characteristics to animals, in this case great apes. Characteristics which usually come under this heading would typically be the ungraspable and unmeasurable aspects of apes, precisely those which, as we saw, can hardly be tackled within the framework of animal science. However, only if an animal scientist goes as far as saying that all animal traits which cannot be measured in the laboratory therefore do not exist, can he or she safely brand any comparison of nonphysical animal features as misplaced anthropomorphism.

Meanwhile, anthropologists and other social scientists jealously guard what they see as the human domain and therefore tend to applaud the biologist's fears of anthropomorphism. What currently tend to be denounced as anthropomorphisms are those characterisations which the anthropologist is keen to reserve for humans and humans alone. The threat of biological determinism tends to be countered with anthropocentrism.

But how can one know how apes are different from or similar to us if one declines to ask the same questions about the two? Presently there are some courageous animal scientists who do say that animals are more human-like and less object-like than their own sciences will have us believe. However, often they will only say such things off the record and rather apologetically, aware that they are committing something close to sacrilege from the point of view of both the biological and the social sciences. As a result, such people frequently reveal a curious tension between the accepted biological codes and their own experiences with animal personhood.

Ape Culture

Nevertheless, some biologists and ethologists have in recent years discovered phenomena among animals – particularly apes – which anthropologists and sociologists would label as belonging to the categories of *society* and *culture*, should they care to study them. Without presuming ape societies to be totally identical with presently existing human societies, it is beyond doubt that apes do live in societies which constitute more than just a random aggregation of individuals. Jane Goodall, Frans de Waal and many other researchers who regularly study groups of chimpanzees, and Dian Fossey, who studied gorillas, have given numerous accounts of apes apparently observing social rules and norms and conveying meaningful messages to each other. Moreover, they can communicate *about* things not present here-and-now. Their communications involve much more than just the uttering of emotions like fear, pain or anger.[15]

Shared meanings, customs and values are part of the culture which backs up the actions and interactions of all those who share a society. Whether conscious or unconscious, explicit or implicit, ideas and meanings come to be created collectively by a social group and are then memorised, transmitted and learned, which is how they travel through time and space.

Not only Goodall and Fossey, well known from studies of African apes in the wild, but also Emil Menzel and Frans de Waal, both of whom studied (semi-) captive groups of chimpanzees, report convincing instances of social standards and social meaning prevailing in ape societies.[16] In addition, there have been many reports on the creation and utilisation of material forms of culture among apes, such as tools. Notably Jane Goodall but also Yukimaru Sugiyama and Geza Teleki have been baffled by the intricacy and complexity of chimpanzee tool-making and tool-using, as well as chimpanzee tool traditions.[17] As far as another cultural expression – language – is concerned, this virtually boils down to the conveying of meaning by way of collectively based symbols. The laboratory experiments devised to teach great apes a number of 'humanised' languages are well known in this context. Whether gestural, plastic-words or computer language, all these enterprises constitute examples of apes learning the language and the meanings prevalent in what was essentially a mixed human–ape society, containing at least one ape and one human instructor.

Forms of culture, language, meanings, tools and abstract thinking have all been encountered in the societies of great apes. It is not for me to go into details, since other contributors to this volume are much more suited for the task. What I would like to emphasise, though, is the need for an anthropological approach to ape societies and cultures. As noted above, all we have at present is an anthropocentric anthropology of humans, who may or may not be studied in their relation to animals. I would like to extend the range of anthropology, so to speak.

Moreover, declaring humans and other great apes to be all of one kind – as belonging to a community of equal subjects – cannot but lead to a total revision of current ape research procedures. The proposed change of attitude calls for a release of all ape objects of study from human laboratories. Instead it seems fitting that humans should ask permission from their ape subjects before intruding upon their societies, just as anthropologists need to do whenever they come upon a foreign community of people. Incidentally, both Fossey and Goodall have made it quite clear that in their respective research situations they initially felt like intruding visitors.

Anthropology is very much the science of the 'other'. Instead of a subject–object approach it possesses a pre-eminently subject–subject method: participant observation of, and living with, people in other societies and other cultures. In contrast to laboratory scientists, who are content to register and measure from without, anthropologists will want to study from within, as much as they possibly can. They will have to immerse themselves in the other's sphere, sharing their people's daily life, learning their language as well as their habits and views. Ideally they will seek to become Indian with the Indians. Participant observation is virtually an exercise in *empathy*.

Admittedly there is a sense in which we can *never* be at one with the other. We will never succeed in leaping over our own socialisation and our Western history. Thus Western anthropologists cannot ever totally know or understand the people they study. Anthropologists necessarily remain prisoners of their own background. In this fundamental sense ethnocentrism can never be totally overcome. But at least as far as human subjects are concerned, the anthropologist is supposed to tread upon this unknowable ground with respect rather than with disdain.

A similar situation is bound to occur while studying ape subjects. No scientist can ever totally transcend his anthropocentrism, in that he cannot leap over his own humanity and the typically human perspective. In that sense our fellow apes remain unknowable.

Curiously, a number of biologists faced with the shortcomings of their own traditional paradigms have already realised anthropology's potential for the study of apes, better than the anthropologists themselves have. Thus the author of *Primate Visions*, Donna Haraway, has in an earlier work already contemplated an 'anthropology of apes'.[18]

What would this mean in real terms? One can only speculate. Instead of making apes conform to human societies, human cultures and human-created language, to determine how apes are social and cultural the anthropologist could go and share the social lives of apes for a given period of time. Instead of imposing our language forms and our social order upon the apes these anthropologists would be required to be taught natural ape language by their chimpanzee, gorilla or orang-utan instructors and obey the rules prevailing in respective chimpanzee, gorilla or orang-utan societies. Ideally, anthropologists studying ape societies must become 'ape with the apes'. They should, for instance, take note of ape world views. Apes tend to see, smell, feel, taste or hear the world against the background of their own frame of reference. They, like us, distinguish and select among sense impressions, distinctions which we may not even know are there. Those features of the world that we would select as salient characteristics need not in any way correspond with what would be relevant to a chimpanzee, gorilla or orang-utan. There is a way things look, taste, smell, feel or sound to an ape, a way of which we will have no idea as long as we insist that the only things worth knowing about are our own constructions of the world.

Would an anthropology of great apes be feasible at all? We won't know until somebody tries. One thing is for sure. Most ape language experiments have been set up primarily to facilitate human (scientific) research. It looks as if it largely was the scientists' own need – due to a defective and limited command of body language on their part – which has dictated the efforts to teach apes human language. In other words, so far the subject–object approach has gone virtually unchallenged: anthropocentrism still prevails.

How about apes teaching humans *their* language? Would we be such good pupils as they are? Hitherto it has been the apes who found themselves at the receiving end of the social and cultural process. It was they who were forced to take a step toward us, humans. It was they who were required to make sense of *human* communication forms and *human* meanings in *human* settings. Incidentally, both Jane Goodall and Dian Fossey have at times attempted to reverse the roles. They actively tried to conform to ape ways, partly in order to reassure individuals, to convince them of their own harmlessness or to prevent attacks upon themselves. However, their overall aim was to study chimp and gorilla society with a minimum of human interference, though Goodall and her staff did provide bananas to the chimps at Gombe.

Contemplating anthropological fieldwork among ape subjects, one may well ask if by doing so one is not destroying precisely that which one seeks to know. Would the work done by a human anthropologist be beneficial at all to the apes? Or would their authentic ape societies in no time be colonised or destroyed by us? Wouldn't it be better to leave them be?

Such questions, hinting at the fact that anthropologists themselves are not acting in a social vacuum, are turning up time and again in circles of anthropologists, but as yet only with respect to human subjects. Again, from both Goodall's and Fossey's work it is apparent that these two scholars have wrestled frequently with similar dilemmas. (Perhaps one could even name them 'animal anthropologists avant la lettre'.) Posing questions of this kind seems even more justified now that we are considering fully fledged anthropological fieldwork among apes. *Homo sapiens* is not exactly known for its fair treatment of the human other, let alone the animal other. But to get an idea of our own reputation throughout the animal kingdom, we would have to be taught an immense number of animal languages and world views. Let's start with the great apes.

Notes

1. See, for instance, the work by Marvin Harris, as well as A. Leeds and A. P. Vayda (eds), *Man, Culture and Animals in Human Ecological Adjustments*, based on a symposium at the Denver Meeting of the American Association for the Advancement of Science on 30 December 1961 (American Association for the Advancement of Science, Washington, 1965).

2. See the work of Claude Lévi-Strauss, Edmund Leach and Mary Douglas. See also Steven Lonsdale, *Animals and the Origins of Dance* (Thames and Hudson, London, 1981).

3. This fact has also been established by Elizabeth A. Lawrence, *Rodeo – An Anthropologist Looks at the Wild and the Tame* (University of Chicago Press, Chicago, 1984), p. 3.

4. See A. L. Kroeber and Clyde Kluckhohn, *Culture, a Critical Review of Concepts and Definitions* (Vintage Books, Random House, New York, 1952).

5. Cf. Ralph L. Holloway Jr, 'Culture, a human domain', in *Current Anthropology*, vol. 10, no. 4 (1969) pp. 395–412; see also Roger M. Keesing, *Cultural Anthropology, A Contemporary Perspective* (Holt, Rinehart and Winston, New York, 1981), pp. 68–9.

6. Paul Bohannan, 'Rethinking culture: a project for current anthropologists', *Current Anthropology*, vol. 14, no. 4 (1973) pp. 357–72.
7. Cf. Nico Wilterdink, 'Biologie en sociologie, argumenten voor een ethologisch gefundeerde sociologie', *Sociologische Gids*, vol. XXII, no. 1 (1975) p. 8.
8. Cf. Charles Woolfson, *The Labour Theory of Culture. A Re-examination of Engels's Theory of Human Origins* (Routledge and Kegan Paul, London, 1982).
9. See, for a discussion of this attitude, Roger Trigg, *The Shaping of Man, Philosophical Aspects of Sociobiology* (Basil Blackwell, Oxford, 1982), pp. 78–102.
10. See Robert Wokler, 'Perfectible apes in decadent cultures, Rousseau's anthropology revisited', *Daedalus, Journal of the American Academy of Arts and Sciences*, vol. 107, no. 3 (1978) p. 110.
11. Barbara Noske, *Humans and Other Animals, Beyond the Boundaries of Anthropology* (Pluto Press, London, 1989).
12. See E.J. Dijksterhuis, *De mechanisering van het wereldbeeld* (Meulenhoff, Amsterdam, 1985; first published 1950).
13. See Adolf Portmann, *Biologie und Geist* (Suhrkamp, Frankfurt am Main, 1973).
14. Michael Ruse, 'Sociobiology: a philosophical analysis', in Arthur L. Caplan (ed.), *The Sociobiology Debate, Readings on the Ethical and Scientific Issues Concerning Sociobiology* (Harper & Row, New York, 1978).
15. See Adrian Desmond, *The Ape's Reflexion* (Quartet Books, London, 1980).
16. See Jane Goodall, *The Chimpanzees of Gombe, Patterns of Behavior* (The Belknap Press of Harvard University Press, Cambridge, MA, 1986); see also Dian Fossey, *Gorillas in the Mist* (Penguin, Harmondsworth, 1985); see furthermore Emil Menzel, 'Human language – who needs it?' in Georgina Ferry (ed.), *The Understanding of Animals* (Basil Blackwell, Oxford, 1984); and Frans de Waal's two books: *Chimpansee politiek, macht en seks bij mensapen* (Chimpanzee Politics) (Becht, Amsterdam, 1982) and *Verzoening, vrede stichten onder apen en mensen* (Peacemaking among Primates) (Het Spectrum, Utrecht, 1988).
17. See Goodall, *The Chimpanzees of Gombe*; see also Y. Sugiyama, referred to by Frans de Waal, 'Het menselijk voetstuk; gedragsovereenkomsten tussen de mens en andere primaten', in F.B.M. de Waal (ed.), *Sociobiologie ter discussie, evolutionaire wortels van menselijk gedrag?* (Bohn, Scheltema & Holkema, Utrecht/Antwerp, 1981); see also Geza Teleki, quoted by Adrian Desmond, *The Ape's Reflexion*, pp. 144–6.
18. Donna Haraway, 'Primatology is politics by other means', in Ruth Bleier (ed.), *Feminist Approaches to Science* (Pergamon Press, New York, 1986).

27

Aping Persons – Pro and Con

STEVE F. SAPONTZIS

Steve Sapontzis is professor of philosophy on the Hayward cam-
pus of the California State University. He is a co-founder and co-
editor of the journal Between the Species, *which publishes articles*
on ethics and our relations with nonhuman animals. In his book
Morals, Reason and Animals, *and in many articles, he has care-*
fully explored a wide range of arguments about the moral status
of animals. His role in this volume is, as he himself notes, that of
the gadfly. He questions a premise shared by many of the other
contributors to this book: that the special intellectual capacities of
the great apes should entitle them to a higher moral status than
other beings with interests.

On reviewing the preliminary list of eminent contributors to this
volume, I had to wonder what I might add that would not be merely
redundant. When I was originally contacted by Paola Cavalieri for a
little article on chimpanzees and 'human rights' for *Etica & Animali*, the
concept of personhood had a prominent place in her concerns, as it does
in much of moral philosophy. That concept is not employed in the
Declaration on Great Apes, but I would still like to say a few words
about it. Discussing personhood will lead us to a presumption – a kind
of intellectual bias – which underlies this focus on great apes and which
I find morally objectionable. Naturally, I will want to say a few words
about that – every volume needs a bit of the gadfly. Next, I have a few
thoughts, of a 'yes-and-no' nature, to contribute to the discussion of
whether our relations with nonhuman great apes should be governed by
the same basic moral principles or rights as govern our relations with
human beings. Having laboured these theoretical points, I will, of
course, want to change gear and conclude with a few political
comments.

Speciesism Revisited – The Intellectual Bias of Persons

To begin with, then, what about 'personhood'? Would it make sense, even though it would certainly sound strange, to say that nonhuman great apes are persons?

'Person' has both a descriptive and an evaluative meaning. In the evaluative sense of the term, 'person' refers to beings whose interests are morally or legally protected against routine exploitation by those whose actions can be directly influenced by moral or legal concepts. Persons are those whom morality or the law indicates we, as moral or legal agents, must treat fairly: must not, as Kant would say, treat as mere means to the satisfaction of our interests.[1] I think – and have argued at length for this conclusion elsewhere[2] – that we should regard all beings with interests (i.e. all beings with feelings) as persons in this evaluative sense of the term. That is, I think we should treat all beings with interests fairly, regarding none of them as mere means to the satisfaction of our interests. That is what animal liberation is all about.

Very briefly, the argument for this conclusion runs as follows. Morality is goal-directed activity which aims at making the world a better place in terms of reduced suffering and frustration, increased happiness and fulfilment, a wider reign of fairness and respect for others, and enhanced presence and effectiveness of such virtues as kindness and impartiality. Through our exploitation of nonhuman animals we detract from all of these moral goals. Factory farming, fur trapping and other exploitations of nonhuman animals increase the suffering and frustration in the world and reduce happiness and fulfilment – the exact opposite of our moral goals. In using our vast power over nonhuman animals to make them bear burdens and suffer losses so that we may be comfortable and prosperous, we extend and enforce a reign of tyranny and disregard, verging on contempt, for others – again, the exact opposite of our moral goals. Finally, by giving revulsion at and compassion for the suffering of nonhuman animals the demeaning labels of 'squeamishness' and 'sentimentality' and by conditioning children to disregard such feelings as they learn to hunt, butcher or vivisect nonhuman animals, we limit and inhibit the virtues of which we are capable – again, just the opposite of our moral goals. Consequently, in all these ways our goal of making the world a morally better place will be more effectively pursued by liberating from human exploitation all those capable of suffering and happiness and of being treated fairly and virtuously.

Nevertheless, there is a strong tendency, even among advocates of animal rights, to retain a close association between 'person' in the evaluative sense and 'person' in the descriptive sense, where it is just another name for human beings. Some writers, such as Tom Regan,[3] suggest that only the more intellectually sophisticated nonhuman animals merit the protection of their interests against human exploitation, and others, such as Peter Singer,[4] maintain that more intellectually sophisticated lives have a higher value than do less intellectually sophisticated lives. It is not surprising that intellectuals retain a bias in favour of the intellectual, but this bias opens the door to critics, such as J. Baird Callicott,[5] who contend that animal rights remains an anthropocentric value system. Instead of being human chauvinists, these critics maintain, animal liberationists are human-like chauvinists, but that represents only a minor change.

Focusing animal rights concern and activity on nonhuman great apes and other nonhuman primates expresses and continues this bias. We are called on to recognise that harmful experiments on nonhuman great apes are wrong because these apes are genetically so much like us or because they are so intelligent, again like us. Such calls clearly retain an anthropocentric view of the world, modifying it only through recognising that we are not an utterly unique life form.

Rejecting our species bias – overcoming speciesism – requires that we also reject our bias in favour of the intellectual (at least as a criterion of the value of life or of personhood in the evaluative sense). Overcoming speciesism requires going beyond the modest extension of our moral horizons to include intellectually sophisticated, nonhuman animals, such as chimpanzees and whales. It requires recognising not only that the origin of value does not lie in anything that is peculiarly human; it also requires recognising that the origin of value does not lie in anything that is human-like or that humans may be assured they have the most of (because they are the most intellectually sophisticated beings around).

An affective value theory can provide the needed foundation for such an unbiased world view. Such a theory holds that values originate with feelings, such as pleasure and pain, fulfilment and frustration, joy and sorrow, excitement and depression, and so forth. Without such feelings there are only matters of fact and definition, i.e. physical and conceptual configurations and changes. Consider a piece of paper you have crumpled and thrown in the wastebasket. It may uncrumple a bit and change its position in the basket. You may even hear this happen, but as long as you don't care about it and as long as this change doesn't impact on the feelings of any other being, there is no value here. But when feelings become involved, these configurations and changes can take on value: they can become contributions to or detractions from a world which is pleasant or painful, fulfilling or frustrating, joyful or sorrowful, exciting or depressing, and so forth. That paper's uncrumpling may flip it out of the basket on to the floor, where you have to pick it up and throw it away again – how annoying for you! The change has acquired value.

Now, feelings are not peculiarly human nor peculiar to human-like animals. Both behavioural and physiological evidence indicate that feelings are part of the psychology and worlds of a wide variety of nonhuman animals, including fish and reptiles as well as birds and mammals. Furthermore, there is no reason to believe that intellectually sophisticated beings have feelings to a quantitatively or qualitatively greater degree than do intellectually unsophisticated beings. Jeremy Bentham, who maintained that all moral values derive from contributions to or detractions from happiness, noted seven dimensions to the value of feelings: intensity, duration, certainty, extent, fecundity, purity and propinquity.[6] So, even if intellectually more sophisticated beings can enjoy a wider variety of feelings, those who are intellectually less sophisticated can compensate for and even overcome this deficit through greater intensity, duration, purity, extent, etc., of their feelings. Next time you go to the beach or the park, take a look around and see who is happiest and enjoying the day to the fullest. Is it the intellectually sophisticated human adults, or is it the children and the dogs?

Consequently, if we recognise that all beings with feelings should be liberated from human exploitation precisely because they are feeling beings, we will have overcome speciesism and freed our morality from anthropocentric prejudice. In such a morality we are called on to recognise not only that the exploitation of human-like animals, such as nonhuman great apes, is wrong (*prima facie*) but also that the exploitation of rats, lizards, fish and any other kind of feeling being, human-like or not, intellectually sophisticated or not, is wrong (*prima facie*).

Rights Revisited – Taking Differences Seriously

Now, if we should treat all feeling beings as persons in that evaluative sense, does it follow that we should treat them as we do human beings? Does it follow that we should extend 'human rights' to nonhuman beings? Is that what would be involved in welcoming nonhuman great apes into the community of equals?

Protecting the interests of nonhuman animals against human exploitation requires extending to them the same basic rights as humans currently (are supposed to) enjoy only if nonhuman animals have the same basic interests as we do and only if extending rights is the appropriate way to secure that protection. It is not obvious that either of these conditions is the case.

Whether nonhuman animals have the same basic interests as we do depends not only on what their and our interests are but also on how they are characterised. We can describe interests in a sufficiently general way, so that all feeling animals can be described as (normally) having the same basic interests as we do. For example, we can say that we all have interests in life, liberty and the pursuit of happiness and, therefore, that we all need the same sorts of protection for our interests. The three principles or rights mentioned in the Declaration on Great Apes – the right to life, the protection of individual liberty and the prohibition of torture – are of this very general sort.

However, we can also describe interests more specifically. This can lead us to conclude that we have interests nonhuman animals do not, and vice versa, and, consequently, that we need protections they do not, and vice versa. For example, part of the pursuit of happiness for many humans is the freedom to pursue their religious beliefs, and they need a right to religious freedom (or some other moral/legal instrument) to protect that interest. Nonhuman animals do not appear to have any such interest and, consequently, do not need that right. Conversely, part of pursuing happiness for some nonhuman animals is being able to stretch their wings; so they need a right (or some other moral/legal instrument) to protect that interest. Lacking wings, we need no such right.

Thus, the answer to whether nonhuman great apes should be extended the same basic moral and legal rights as humans depends in part on whether these basic rights are being formulated in a general or specific manner. In developing moral and legal codes which people would be supposed to follow and to which they could be held accountable, specific formulations would have to be employed. Consequently, at this level the answer must be 'no': even in thoroughly non-speciesist, animal-respecting moral and legal codes, nonhuman great apes need not have the same basic moral and legal rights as humans.

And vice versa, let us not forget. There is a tendency to think that if we conclude that nonhuman animals are not to enjoy all the rights of humans, it is because they are entitled only to a few of those rights. However, basing moral and legal protections on specific interests can also lead to the conclusion that nonhuman animals should have rights that humans do not need. So, specific nonhuman and human rights can be different without the former being merely a subgroup of the latter, and, consequently, without suggesting that the nonhumans are morally or legally less worthy beings.

However, the development of these specific, animal-respecting codes would be directed by those more general formulations of rights, such as extending to nonhuman great apes the same sorts of moral and legal protection of their interests in life, liberty and the pursuit of happiness that humans currently (are supposed to) enjoy. So, at this level of principle, the answer to our question could still be 'yes'. In the Declaration on Great Apes the idea of a community of equals is defined at this very general level, where the often misleading claim of, or call for, human/nonhuman 'equality' makes sense.

Still, I say only that the answer to our question about extending basic human *rights* to nonhuman animals 'could be' yes, for there are still other complications determining the final answer to that question. For example, when discussing the interests of nonhuman animals we ordinarily focus on things that these animals can take an interest in, such as food and exercise. But in addition to these, there are things that nonhuman animals cannot take an interest in but in which they none the less have an interest, since these things impact on their feelings. Consequently, developing moral and legal codes to protect the interests of all feeling animals would require considering things in which nonhuman animals have an interest even though they cannot take an interest in them.

Voting is an example of this. Nonhuman animals cannot understand what voting is all about and how it affects their interests. Consequently, unlike humans, nonhuman animals do not feel vulnerable or demeaned because they are not allowed to vote. None the less, which politicians are elected and which are not can critically affect their interests. For instance, it would benefit the interests of nonhuman great apes if politicians who oppose harmful experiments on nonhuman primates were elected. Thus, nonhuman great apes have an interest in voting, even though they cannot take an interest in voting. So, if we are to extend to nonhuman great apes the same sorts of moral and legal protections of their interests that humans currently (are supposed to) enjoy, then this interest in voting must enter into our deliberations.

We might conclude that nonhuman animals need the right to vote – through a concerned, informed guardian – in order to protect their interests in life, liberty and the pursuit of happiness. However, the difficulties of implementing such a right are so great as to render that conclusion thoroughly implausible. How are nonhuman animals to be counted and registered, and how are human proxy voters to be selected and nonhuman proxies assigned to them? Also, in the case of children, whose interests are also affected by voting, we do not conclude that protecting their interests entails that they have the right to vote. By analogy, protecting the interests of nonhuman great apes would not entail such a right.

Such cases indicate that it is simplistic to infer that because something has an impact on the basic interests of members of a group, and those interests should be protected, we must conclude that members of that group have a right to (or against) that something. There is a tendency, especially in the United States, immediately – and vociferously – to employ the concept of rights whenever questions of protecting interests arise. But that concept does not readily fit all such situations, especially when the interests in question are not those of normal, human adults, i.e. those of intellectually sophisticated, autonomous agents. Consequently, morally and legally protecting the basic interests of nonhuman animals may involve some ingenuity and thoughtful working with a variety of moral and legal categories, rather than automatically demanding rights for nonhuman animals to (or against) those things which (can, will, would) have an impact on their basic interests. For example, the Declaration on Great Apes defines the community of equals in terms of 'moral principles or rights', and among the three principles enumerated, only one is identified as a right, the others being a 'protection' and a 'prohibition'.

In developing and deploying these categories, however, it must be clearly understood that they afford the same level of protection for the interests of those who are not intellectually sophisticated, moral and legal agents as rights provide for the interests of such agents. That is, when the interests of a being protected by one of these categories conflicts with the interests of an agent protected by a right, the right cannot automatically override the other category. To avoid the sham protection of our contemporary 'humane' values and laws concerning nonhuman animals – a protection that is easily overridden even by the trivial desires to eat pale veal and to save a few pennies on a dozen eggs – the moral and legal categories to be developed and deployed in liberating nonhuman animals from human exploitation must share the exalted status that only the concept of rights currently commands. This is another aspect of overcoming the intellectual bias in our speciesism: currently our most powerful moral and legal concept, 'rights', is one which is suited to the capacities and conditions of intellectually sophisticated agents; in a liberated ethic, concepts suited to the capacities and conditions of feeling beings who are not intellectually sophisticated agents must enjoy equal status and power with the concept of rights.

To summarise, while nonhuman great apes should be persons in the evaluative sense of the term – which is to say that they should enjoy the same level of moral and legal protection of their interests as humans do (or are supposed to) – this protection need not take the form of assigning rights in every case. Thus, liberating nonhuman great apes from human exploitation need not take the form of extending 'human rights' to them. These apes will not need some of the rights humans do, if they do not share in all human interests, but they may also need some rights that we do not, if they have interests which we do not share. Also, other moral and legal protective categories may be more appropriate than rights to the capabilities and conditions of these apes. Finally, from the perspective of liberation moral theory, nonhuman great apes do not obviously have any more claim on personhood and this protection of interests than do other, less intellectually sophisticated, nonhuman animals.

Liberation Revisited – The Real Pursuit of Ideals

None the less, from the perspective of liberation moral practice it may be appropriate and even politically astute to emphasise the human-like characteristics of nonhuman great apes and to seek the moral and legal protection of their interests as persons before seeking such protection of interests for all feeling animals.

We humans have social instincts: we tend to divide up the world into 'us' and 'them' and to feel much more strongly obligated to those whom we consider kin. So, to the extent that we can bring people to recognise that nonhuman great apes are members of our biological 'family' and can thereby bring people to extend their fellow-feelings to embrace these extended family members, we are more likely to secure for nonhuman great apes the protection of their interests against human exploitation that they morally deserve and desperately need. In this way there may be a practical, political pride of place for nonhuman great apes – similar to that for companion animals, who are members of our socially extended families – even though ultimately, without reference to human instincts and propensities, there is theoretically no obvious pride of place for them, or for any other feeling species.

This practical conclusion should not be condemned as a compromise of liberation ideals. Too often when doing moral philosophy we forget that it is supposed to be a practical science, i.e. a study whose conclusions are not theories but actions. Ideals are needed to guide moral action, but we cannot deduce what is to be done from ideals alone. In addition to ideals, action is determined by the material with which we have to work to realise those ideals. And the material for animal liberation – as for all moral change – is human beings as they currently are, with their native (in)capacities and (in)sensitivities, established cultures, contemporary (im)moral beliefs and practices, current economic dependencies and present world views. Developing and deploying concepts and arguments which will move people as they are to make the world a better place is the proper conclusion of moral philosophy, and moving them to make the world a better place for nonhuman animals is the proper conclusion of animal liberation philosophy. Developing moral theory and ideals, as has been done in this chapter, is only a means to that end.

Ideals must be kept in view if our efforts for nonhuman animals are not to be co-opted and to effect merely rhetorical, complacent changes – as when vivisectors now readily agree that nonhuman animals have rights but then go on to assert that those rights are respected in humane laboratory sacrifices of nonhuman animals. On the other hand, those who insist that all animal liberation projects focus exclusively on the ideal, and disdainfully reject all accommodation of liberation ideals to current realities, will likely succeed only in feeling that their hands are clean and their consciences are pure. Wilfully out of touch with many of the forces that move and shape reality, they are not likely to succeed in helping nonhuman animals, and their cherished, beautiful ideals will likely remain mere ideals while nonhuman animals continue to suffer and die without relief.

So, engaging in campaigns – such as this one to extend protective moral and legal principles and rights to nonhuman great apes – which take advantage of anthropocentrism and other human imperfections and which, consequently, fall short of the ideals of animal liberation, is not compromising those ideals. It is implementing and pursuing those ideals in the world as it is. That, rather than theoretical precision and purity of conscience, is what moral philosophy and animal activism are finally all about.

Notes

1. 'Act so that you treat humanity, whether in your own person or in that of another, always as an end and never as a means only.' Immanuel Kant, *Foundations of the Metaphysics of Morals*, trans. Lewis White Beck (The Library of Liberal Arts, Indianapolis, 1959), p. 47.
2. S.F. Sapontzis, *Morals, Reason, and Animals* (Temple University Press, Philadelphia, 1987).
3. T. Regan, *The Case for Animal Rights* (University of California Press, Berkeley, 1983).
4. P. Singer, *Animal Liberation: A New Ethics for Our Treatment of Animals* (New York Review of Books, New York, 1975/1990).
5. J. Baird Callicott, *In Defence of the Land Ethic* (State University of New York Press, Buffalo, 1989).
6. J. Bentham, *An Introduction to the Principles of Morals and Legislation* (1789), chapter IV.

VI

Reality

28

Items of Property

DAVID CANTOR

David Cantor is information specialist in the Research and Invest-
igations Department at People for the Ethical Treatment of
Animals, in Washington, DC. He has written many articles on the
use of animals in schools, and on other forms of animal exploita-
tion. The following essay provides examples of how the present
legal and moral status of the great apes permits people to behave
toward them.

A few thousand great apes currently live in the United States. Some 2,000 chimpanzees are in laboratories, 800–900 in zoos, and a few in entertainment. Ten to twenty orang-utans are used for entertainment, fifteen to twenty in laboratories, and several hundred are kept in zoos. Almost 300 gorillas are in zoos, ten to fifteen in laboratories, and currently none are known to be used for entertainment, though one is kept on display in a shopping centre in Tacoma, Washington.

Two main factors have prevented the numbers of great apes in laboratories from increasing significantly in recent years: their low rate of reproduction in captivity and the Convention on International Trade in Endangered Species of Wild Fauna and Flora (CITES). CITES, based in Switzerland, oversees a trade treaty signed by 111 nations, including the UK and the USA. It has drastically diminished the export of many species – including great apes – from their native habitats, though illegal shipments still occur. This leaves to experimenters seeking protected animals those smuggled into the country or those born in captivity.

Chimpanzees in a Laboratory

In 1987, a group called True Friends entered SEMA (formerly Meloy), a laboratory in Rockville, Maryland, where chimpanzees and many other primates were known to be kept for experiments under contract with the National Institutes of Health (NIH). True Friends photographed and videotaped the laboratory and removed four chimpanzees, providing concrete evidence of conditions under which animals lived and died at SEMA. People for the Ethical Treatment of Animals (PETA) made the facts public, adding to its report information obtained from government documents and other written sources. Inasmuch as standards of care for animals in laboratories are determined by the same criteria throughout the USA, and government inspections are conducted on the same basis and by the same agencies at all facilities, the story of SEMA can be considered broadly representative of other US research laboratories, but particulars must always be documented to be considered accurate.

At the time of the True Friends' raid on SEMA, the company was infecting many kinds of primates with influenza, hepatitis and other diseases, as well as giving them cancer and infecting them with HIV, the human AIDS virus. Chimpanzees were infected with hepatitis and HIV. Experimenters would record symptoms and the course of the illnesses and test possible treatments.

Nearly 700 primates were living alone in isolettes – steel cages designed for one animal, devoid of anything that could provide stimulation. Cages with front doors of metal mesh or bars had solid sides, preventing animals living side by side from seeing each other. Large rooms were filled with these isolettes, so the animals could hear each other and in some cases could see each other, but could not socialise. Infected with diseases, animals spent their entire lives in this way. A chimpanzee conceivably might live fifty years in an isolette, leaving it only when experimental procedures or cage maintenance required, or after his or her death.

According to the contract, SEMA experimenters were to carry out NIH researchers' protocols and were 'to house and maintain nonhuman primates while conducting directed AIDS studies and studies of various enteric and respiratory diseases'. This same contract, executed on 18 December 1984, stated:

> The Contractor shall provide isolated care and housing for chimpanzees (or other animal models if directed by the Contracting Officer) to be used for the study of Acquired Immunodeficiency Syndrome (AIDS). All animals to be used for this project shall be housed in one room separate and distinct from all other animals.

It further states:

> Specifically, the Contractor shall:
> (1) Provide an animal holding facility capable of holding approximately nine chimpanzees (5 of which weighing up to

25 kilograms and 4 of which weighing up to 20 kilograms
. . .) for AIDS research.
(2) Hold these animals in cages and isolator units which will be
provided by the Government.
(3) Treat or infect, at the direction of the Project Officer, the
animals with the material supplied and obtain bleedings,
biopsies, perform laparotomies, and other specimens which
will be analyzed by NIH scientists for existence of infection.
(The Government estimates the following on an animal basis:
500 bleedings [plasmaphereses/leukophereses]; 100 biopsies;
and 100 laparotomies).[1]

(The contract also described experiments using squirrel monkeys,
rhesus monkeys and other primates.)

One experiment contracted by the National Institute of Allergies and
Infectious Diseases (NIAID, a branch of NIH) called for SEMA to
inoculate up to ten chimpanzees each year, indefinitely, with HTLV-III
or other newly recognised human retroviruses and follow them for
evidence of infection.[2]

Like any laboratory under contract with NIH, SEMA is required to
maintain accreditation with the American Association for Accreditation
of Laboratory Animal Care (AAALAC), a non-profit-making organisation, by adhering to US Public Health Service (PHS) guidelines for
animal care. PHS guidelines state:

The housing system [for animals used in research] should:
• provide space that is adequate, permits freedom of movement and normal postural adjustments, and has a resting
place appropriate to the species;
• provide a comfortable environment;
• provide an escape-proof enclosure that confines animals
safely;
• provide easy access to food and water;
• provide adequate ventilation;
• meet the biological needs of the animals, e.g., maintenance of
body temperature, urination, defecation, and, if appropriate,
reproduction;
• keep the animals dry and clean, consistent with species
requirements;
• avoid unnecessary physical restraint; and
• protect the animals from known hazards.[3]

For chimpanzees weighing more than 25 kilograms (55 pounds), the
Guide recommends a minimum cage size of 2.33 square metres (25.1

square feet) of floor area for each animal and a minimum height of 2.13 metres (84 inches). For chimpanzees of 15–25 kilograms (33–55 pounds), the minimum is 0.74 square metres (8 square feet) per animal, with a height of 91.44 centimetres (36 inches). Thus, in the contract quoted above, calling for some chimpanzees weighing more than 25 kilograms and some weighing less, a chimpanzee weighing 24 kilograms (53 pounds) could be kept for years in a cage one-third the area of that inhabited by a chimpanzee only slightly larger at 26 kilograms (57 pounds).[4]

Like all US research facilities using animals, SEMA is also governed by the federal Animal Welfare Act, which since 1966 has required the US Department of Agriculture (USDA), through its Animal and Plant Health Inspection Service (APHIS), to ensure that animals used in experiments, in exhibitions or as 'pets' receive humane care and treatment.[5]

In the two years before PETA publicised information about SEMA (1985–6), APHIS inspection reports showed nineteen violations of the Animal Welfare Act in fifteen categories, including space requirements, feeding, watering, cleaning, housekeeping, veterinary care, pest control, employee training, drainage, interior surfaces, lighting and ventilation. All five inspection reports over those two years stated that primates were 'too large for their cages'. With regard to cleanliness, an inspection report stated:

> Many primary enclosures had . . . excreta caked onto the bars, accumulation of dust and food particles . . . caked food and debris along rubber gaskets of the chamber doors. All dark corners inspected revealed roaches and mouse feces; evidence of vermin was everywhere.

Feeding receptacles were described as 'grossly contaminated'. One report said, 'Given the . . . deficiencies, it is imperative that both more personnel and better trained personnel be hired to correct them.'[6]

In the same two-year period, sixteen failures to meet PHS guidelines were reported by AAALAC inspectors. As a result, SEMA was placed on probationary status. Between 1981 and 1986, SEMA was on probation almost three times more often than it was accredited. Deficiencies in 1985–6 included confining chimpanzees to cages which did not conform to Animal Welfare Act and AAALAC requirements; some were in cages 40 inches high, 26 inches wide, and 31 inches deep – smaller than the minimum standard for *any* chimpanzee. Faecal material was caked on to bars and sides of cages, and there was excessive urine build-up on suspended waste troughs. Many animals were in need of veterinary care; their symptoms included shivering, skin abrasions, hair loss and vomiting. One monkey was found lying dead on his cage floor.

From 1981 to 1984, SEMA had an extremely high 'accidental' death toll of seventy-eight animals, including five chimpanzees. A chimpanzee known as No. A51 choked on his own vomit; SEMA attributed this to 'a breakdown of normal procedures'. An NIH official said that 'events that led to the death of [chimpanzee] A117', who suffocated in a cage that was too small, 'can and must be prevented'. One chimpanzee died because he was unable to breathe after being anaesthetised with ketamine. The attending veterinarian said that because of increased workloads and too small a staff, technicians were anaesthetising more than one chimpanzee at a time. Consequently, each animal was not adequately monitored to ensure his or her safety. The chairman of the Council on Accreditation of AAALAC wrote in March 1986, 'Presently, primates that die from unsuspected cause, unrelated to experimental protocol, are necropsied . . . Council is concerned, however, that detailed histopathological or microbiological procedures are not undertaken to establish a definitive diagnosis.'[7]

As clearly shown in the video footage obtained at SEMA by True Friends, many of the animals living in isolettes there had become psychotic. Some continually spun in circles and did not react to the presence of visitors. An older male chimpanzee, identified as No.1164, was crouched on his cage floor, rocking back and forth and mumbling incessantly.[8]

In September 1992, the president of SEMA, now called Bioqual, allowed one area of the facility, a kind of 'chimpanzee showcase', to be visited and photographed. This consisted of a continuous series of partitioned kennel-style runs with Plexiglass sides. The chimpanzees in these individual enclosures – there were about twenty of them – could see other chimpanzees in front of, behind and alongside themselves. There was a playroom containing toys and climbing apparatus. In their individual enclosures, the chimpanzees had something to swing on, a few toys and an elevated sleeping platform, and the visitors were told that the chimpanzees received two pieces of fruit each day. The president stated that although hepatitis, influenza and other infectious disease experiments are still carried out at Bioqual, chimpanzees are no longer used in AIDS studies there.[9]

The Orang-utan Trade

Only about 20,000 orang-utans still live freely on the islands of Borneo and Sumatra. Orang-utans are highly prized by zoos, circuses, animal trainers in the entertainment business and wealthy private collectors.

Because of their rarity, orang-utan babies fetch a high price – as high as $50,000 in the United States.[10] Trappers usually kill the mothers – and sometimes other adults and babies – to obtain one young orang-utan. Taking into account the high mortality rate suffered by captured animals, animal rights advocates estimate that certainly two or three, and perhaps as many as ten, animals die for each one who survives the long journey to a zoo or other destination.[11]

The case of one group of captured orang-utans, known as the 'Bangkok Six', has focused public attention on the international primate trade. The six orang-utans were transported without food or water from Borneo to Singapore to Bangkok, Thailand, stuffed into two wooden crates marked 'birds'. The crates, their lids nailed shut, had only pencil-diameter holes for ventilation. One box, carrying three orang-utans, was shipped upside-down. Had officials at Bangkok airport not become curious enough about the un-bird-like cries coming from the crates to X-ray them, the animals would have gone to Belgrade, Yugoslavia, and from there possibly to a Moscow zoo.

The shipment, marked 'personal baggage', also included two sia-mangs, who were presumably to go to Belgrade Zoo as payment for overseeing the orang-utans' passage to Moscow. Although Thailand is a member of CITES, in 1990 it allowed the import and export of non-native species.[12] In fact, officials at Bangkok airport seized the orang-utans and siamangs not because of their endangered status, but because they were labelled incorrectly.

Kurt Schafer, a German resident of Thailand to whom the 'personal baggage' belonged, was charged with no offences by Thai officials, but he was charged in Singapore with exporting the orang-utans and siamangs without official documentation. He voluntarily flew to Singapore, where he was fined just $1,200.[13] The fine in Singapore for failing to flush a public toilet is $100.[14] (In Germany, Schafer was sentenced to seven weeks in jail,[15] though he could have gone to jail for five years for smuggling orang-utans.[16]) Singapore officials cited three 'mitigating factors' in Schafer's case: he was 'do[ing] a friend a favour,' as he told Singapore officials, who then failed to ask the name of the 'friend'; he returned to Singapore voluntarily for legal processing; and the orang-utans and siamangs had been confiscated.[17]

Volunteers from the Wildlife Fund of Thailand named the orang-utans Bambi, Bimbo, Fossey, Ollie, Tanya and Thomas. All were suffering from pneumonia, dehydration and parasitic infestations.[18] Ollie was not expected to live when the International Primate Protection League (IPPL) and the Orang-utan Foundation sent Diana Taylor-Snow, an experienced orang-utan care-giver, to help tend to the orang-utans and prepare them for release into the wild. Ollie improved, but Bimbo, who had liver and spleen damage (possibly caused by having been shipped upside-down from Singapore), was also plagued by frequent sneezes, a runny nose, diarrhoea and ringworm. Despite his physical problems, Bimbo was curious and vivacious. He would make the box he was in move across the floor by leaping around inside it, and, by observing his surroundings, quickly learned how to reach what he wanted.[19]

Taylor-Snow and Wildlife Fund of Thailand volunteers nursed the orang-utans back to the best health possible, given their serious condition, then flew with them back to their home in the forests of Borneo to release them.

In the meantime, the story of the Bangkok Six shifted from the wilds of Borneo to a courtroom in Miami, Florida, where, on 19 February 1992, Matthew Block, head of Worldwide Primates, was indicted by a Miami grand jury for his alleged role in the shipment of the orang-utans, which constituted a violation of CITES laws and the US Endangered Species Act. Block was thought to be the 'friend' for whom Kurt Schafer was doing a 'favour' by accompanying the Bangkok Six shipment in February 1990.

Born in 1961, Matthew Block started trading in exotic birds when he was 13, and, while still young, he became a major figure in the animal import world. By the time he was 21, his exotic bird business was bringing in over $600,000 a year, but stiffer government regulations and more competition persuaded Block to shift to primates and other animals, including elephants, sought by zoos. By 1989, his $1.2-million-a-year enterprise was importing 2,000 monkeys a year[20] and, with about a quarter of the US market,[21] was one of the top three suppliers of nonhuman primates for experimentation.[22]

Block, steadfastly maintaining his innocence, was to go on trial for the Bangkok Six case on 24 August 1992, but on 24 August Hurricane Andrew battered southern Florida and the trial was postponed. Delays caused by the hurricane clogged Miami's legal dockets; meanwhile, Block's attorneys worked out a plea bargain: Block would plead guilty to two misdemeanor charges if the two felony charges were dropped. News of the plea bargain initially cheered the world's animal traders, but in December 1992 the presiding judge, James Kehoe, citing several hundred letters from all over the world expressing concern that justice be served, refused to accept the plea bargain, and a trial date was set for March 1993.

On 9 February 1993, almost three years to the day since the fateful shipment of the Bangkok Six, amid allegations by prosecutors that Block had been involved in a conspiracy with the KGB to smuggle the orang-utans into Russia, Matthew Block pleaded guilty. He will be sentenced on 15 April; the maximum punishment is five years in prison and a fine of $250,000.[23]

Only two of the baby orang-utans survived.[24] 'Ollie and Bimbo died in my arms,' says Taylor-Snow, explaining that it was hard to believe they were alive when she found them. 'They were grey,' rather than the normal deep red-brown of orang-utans. She later heard that Fossey and Thomas died, succumbing to the permanent damage sustained by any infant deprived of food, water and air as were the Bangkok Six.[25]

As long as the orang-utan trade continues, this can happen to any who are unfortunate enough to be captured and taken from their mothers in the forests of Borneo and Sumatra.

Gorillas in Zoos

Money for research involving gorillas became scarce in the late 1970s, the cost of maintaining them being approximately five times that of using the same number of monkeys. Since chimpanzees were classified as a 'threatened' rather than an 'endangered' species like gorillas and orang-utans, chimpanzees remained in labs, but most gorillas and orang-utans were transferred to zoos. Yerkes, by 1988 the only primate research centre keeping gorillas, planned to place all of its gorillas and orang-utans in zoos by 1990, though a few of each species remain at Yerkes.

Timmy, a silverbacked lowland gorilla, began living at the Cleveland Metroparks Zoo in 1966. Isolated from other gorillas for thirty years, he did not get along well with two females to whom he was introduced, and did not mate with them. In 1990, however, the zoo brought into Timmy's enclosure a female slightly older than Timmy: Katie, a.k.a. Kribe Kate. Timmy and Katie quickly began to display affection for each other, playing together, having sexual relations and sleeping in each other's arms.

The Metroparks Zoo is a member of the Gorilla Species Survival Plan (SSP), initiated by the American Association of Zoological Parks and Aquariums (AAZPA) in 1983 'to optimize captive reproduction in North American zoos'.[26] It was hoped that Timmy and Katie would produce offspring, but it turned out that Katie, who had given birth in the past, now had a blocked Fallopian tube and was unable to conceive. The zoo made plans to move Timmy to the Bronx Zoo in New York. This created a conflict between those who believed Timmy should be treated primarily as part of the SSP – as a representative of his species, whose population in the wild is decreasing because human beings are eating them and destroying their habitat – and those who wished to see him respected as an individual gorilla with a strong attachment to another. Timmy was shown to have a very low sperm count, decreasing his chances of reproducing.

Steve Gove, a keeper in the Metroparks Zoo Cat and Primate Building who had worked with Timmy for eighteen years at the time the controversy arose, opposed moving Timmy to New York:

> Timmy is not a very adaptable gorilla, he's proven that. It has taken him years to find a mate he was comfortable with. My biggest fear is that the stress of the move could trigger a heart attack or a stroke. We also worry that the other gorillas might not accept him and [might] hurt him.
>
> The least thing that could happen is that the trauma might send Timmy right back into his shell. He's been very shy since 1966. It was only when Katie got here that he opened up. He could very easily go back to the way he was. None of us want to see that happen.

Zoo director Steve Taylor, who wanted the move to go forward, reportedly said,

> It sickens me when people start to put human emotions in animals. And it demeans the animal. We can't think of them as some kind of magnificent human being, they are animals. When people start saying animals have emotions, they cross the bridge of reality.[27]

The general director of the New York Zoological Society, which includes the Bronx Zoo, wrote in a letter to a newspaper that gorillas are not monogamous and that 'Timmy evolved to manage a harem'.[28]

The California-based animal rights organisation In Defense of Animals, the Network for Ohio Animal Action, and the Animal Protection League (Ohio) organised to prevent the move. More than 1,500 people signed petitions in opposition to it. Demonstrations were held. Attorney Gloria Rowland-Homolak filed for a court order to prevent the Metroparks Zoo shipping Timmy to the Bronx Zoo, and proceedings delayed the move. But in November 1991, Timmy was sedated and put in a crate on a truck bound for New York, as the SSP had intended.

After Timmy was taken away from Metroparks Zoo, a gorilla named Oscar was introduced into Katie's enclosure. The two gorillas fought. Oscar bit Katie's toe, which was then amputated, and Katie was also treated for a bruised back. Eventually, Metroparks Zoo agreed to transfer Katie to another zoo to prevent further attacks by Oscar, who is said to have had a previous history of hostile behaviour in a Kansas zoo. In November 1992, Katie was shipped to Texas in a crate aboard a train. She now lives in the Fort Worth Zoological Park.

Days after the initial brawl between Katie and Oscar, it was reported that Timmy was not mating with the females at the Bronx Zoo, and James Doherty, general curator of the New York Zoological Society, was quoted as saying, 'We want the females to be bred. No one wants to see baby gorillas more than we do,' and that if the ape didn't 'work out', 'we won't keep Timmy indefinitely'.[29]

It was seven months, May 1992, before Timmy was seen mating with a female gorilla at the Bronx Zoo. As of November 1992, he was not confirmed to have impregnated any females despite several matings. Usually it takes several oestrus cycles before a female gorilla becomes pregnant, but Timmy's low sperm count may make offspring all the more unlikely. Meanwhile, an animal protectionist who has monitored events in the lives of Timmy and Katie wrote,

I was at the Bronx Zoo in September [1992] and saw Timmy on two occasions. On both occasions, for the better part of eight hours, Timmy sat on a rock. The only time he moved off of it was to sit against a wall for about twenty minutes ... I was in Cleveland prior to going to New York, and while in Cleveland, I spoke to five people who had also been to the Bronx Zoo, and their sightings were the same as mine.[30]

Acknowledgement – The author wishes to thank PETA senior writer Christine Jackson for drafting the second section.

Notes

1. United States Department of Health and Human Services, Contract No. NO1-AI-52566, 18 December, 1984.
2. United States Department of Health and Human Services, National Institute of Allergies and Infectious Diseases, Animal Study Form (NIH 79–6), 1 October 1986.
3. United States Department of Health and Human Services, *Guide for the Care and Use of Laboratory Animals* (Rev. 1985), pp. 11–12.
4. Ibid., p. 14.
5. United States Code of Federal Regulations, Title 9, Subchapter A.
6. United States Department of Agriculture, Animal and Plant Health Inspection Service, Inspection of Animal Facilities, Sites or Premises, Form 18–8, 9 August 1989. See also People for the Ethical Treatment of Animals (PETA), 'Investigative Report: SEMA Laboratory, Rockville, Maryland' (1986).
7. PETA, 'Investigative Report'.
8. PETA, 'Breaking Barriers' (videotape) (1987).
9. Ingrid E. Newkirk, letter to Dr Jane Goodall, 29 September 1992.

10. William Labbee, 'The primate debate', *New Times* (Miami's News and Arts Weekly), 20–6 November 1991, p. 21.
11. 'Judge refuses plea bargain in Matthew Block case', *AWI Quarterly*, Fall 1992.
12. 'More about the "Bangkok Six"', *IPPL Newsletter*, August 1990, p. 5.
13. 'Background on the "Bangkok Six"', *IPPL Newsletter*, August 1990, p. 6.
14. '"Punishment" – Singapore style', *IPPL Newsletter*, August 1990, p. 6
15. 'Background on the "Bangkok Six"', p. 8.
16. 'More about the "Bangkok Six"', p. 6.
17. '"Punishment" – Singapore style', p. 6.
18. Shirley McGreal, 'Matthew Block indicted on four counts of the Endangered Species Act', *Animal Welfare Institute Quarterly*, Spring 1992.
19. Dianne Taylor-Snow, 'Remembering Bimbo', *IPPL Newsletter*, December 1991, pp. 9–10.
20. Labbee, 'The primate debate', p. 22.
21. Albert Stern, 'Matthew Block animal importer', *New Miami*, April 1989, pp. 40–43.
22. Kathy Glasgow, 'Dead on arrival', *New Times*, 14–20 October 1992, p. 6.
23. Stern, 'Matthew Block', p. 41.
24. 'Miami man admits a plot to ship apes to Eastern Europe', *New York Times*, 10 February 1993.
25. Dianne Taylor-Snow, conversation with the author, 19 February 1993.
26. American Association of Zoological Parks and Aquariums (AAZPA), *Annual Report on Conservation and Science* (1990–1).
27. Michael Sangiacomo, 'Zoo to end love affair, hopes gorilla will mate', *Cleveland Plain Dealer*, 3 September 1991.
28. William G. Conway, 'Bronx gorillas waiting', *Cleveland Plain Dealer*, 2 October 1991.
29. Roberta White, 'Timpotent! Zoo's big ape is just a shy guy', *New York Post*, 24 February 1992.
30. Florence Semon, personal communication, 10 November 1992.

29

The Chimp Farm

BETSY SWART

Betsy Swart is the director of the Washington, DC, office of Friends of Animals. She has written and spoken on a wide variety of animal protection and nonviolence issues and is a doctoral candidate at the University of Maryland. Here she describes a visit to a place called the Chimp Farm, in Florida. It is one of more than 1,200 roadside zoos that exist across the United States. This chapter is an edited version of an article that first appeared in ActionLine, *the Friends of Animals magazine.*

As we paid our $2.25 and walked into Noell's Ark – the Chimp Farm – we passed a woman with three small children. They were standing in front of a barren concrete cage, staring at the lonely chimp inside.

'Hey, you,' yelled one of the children to the chimp. 'Want a peanut? Then DO something.'

'Can't you dance?' yelled another.

The kids giggled and jumped up and down – mimicking chimpanzee movements. But the chimp inside the cage remained silent. He just rocked quietly back and forth – hitting his head very softly against the concrete cage wall. His was a posture of total despair and loneliness.

The woman was about to take her children on a tour of the Farm. I wondered why she would expose her children to such sadness.

'Do you think the animals are happy?' we asked her, gesturing to the long rows of cages containing chimps, monkeys and other animals.

'Oh, of *course*, they're happy,' she said to me disdainfully. 'What kind of question is that? They're only animals.'

Then she and the children began their tour, past the long, desolate rows of pitiful cages.

Noell's Ark is one of the USA's most notorious roadside zoos. It is owned and presided over by Bob and Mae Noell. The couple got their 'show business' start on the vaudeville stage but soon decided that other sorts of entertainment would be more lucrative for them.

In 1939, they attended the World's Fair. And it was there that Bob first got the idea of using animals in their 'act', which in those days was a travelling medicine show catering to rural eastern and southern populations. Soon after the Fair, he bought a gorilla from a dealer and travelled around the country with this new 'member' of their family. The gorilla was a crowd-pleaser, especially after Bob latched on to the idea of having the gorilla 'box' with humans.

Bob and Mae realised they had a hit on their hands. They began acquiring other animals and eventually breeding them.

Soon they expanded the boxing matches to include chimps as well and travelled around the country pocketing a pretty penny from these make-believe bouts. The chimps they used as 'boxers' were, of course, chained and harnessed and otherwise restrained. But many a human 'he-man' would put his money down to try his strength against a chimpanzee – especially when he knew he couldn't lose.

The Noells' travelling days are over now. But the animals which they acquired during their working years – and many, many more – reside in a roadside concrete jungle near Tarpon Springs, Florida. The Chimp Farm is visited by hundreds of tourists each year. Bob and Mae proudly greet visitors and collect admission fees at the door.

The entrance to the Chimp Farm has a carnival-like appearance with huge signs announcing 'REAL LIVE' chimps, gorillas and alligators. There are even cartoon-like plywood cut-outs of various animals outside the Farm's main entrance. Tourists are encouraged to put their faces in the cut-out holes where animal faces would be and then have their pictures taken. 'See how *you* would look as a gorilla,' one of the staff calls out to a child climbing up the exhibit.

The same carnival-like attitude toward animals is encouraged on the inside of the Chimp Farm as well. Food in the form of monkey chow and nuts is available at the front desk or through bubble-gum dispensers. People are encouraged to feed the animals. Unfortunately, most feed them anything that they can manage to pitch into the cages and the bored and stressed animals accept the food gladly. There are a few signs posted on the premises to limit tourists' aggressive behaviour, but neither the Noells nor their staff keep a watchful eye out for the animals' health or safety.

During my two visits, I saw tourists give the animals gum, candy, cookies, Coke and even cigarettes.

But it is easy to see why these 'treats' are welcomed by the animals. They live in the most barren and squalid conditions imaginable. They have nothing to play with – except perhaps their own faeces. Their cages are bare concrete and they are not allowed anything that would interfere with the Noells' clean-up operations, which consist merely of hosing down the cages with the animals inside.

Even a local grocer who offered to donate fruits and vegetables to the animals was refused because these foods make a 'mess' and were deemed inappropriate by the Noells. Except for an occasional tyre hanging from a cage top, there are no psychological enhancements in the cages. Space is minimal. The only animals who are housed together are those the Noells hope to breed. There is no room to exercise for any of the animals; there is no bedding to sleep on; and, in some cases, there is only minimal shelter from the weather.

As I walked along the concrete paths on my tour, I tried to piece together the different animals' histories.

At the entrance to the Farm – as though to entice visitors inside – is a baby chimp in a plexiglass and plywood box. A sign taped to the plexiglass cage proclaims that this baby is the fifty-eighth baby chimp born at the Noells' facility.

The baby spends his day pacing and banging on the walls of his prison. Sometimes he will put his head against the plexiglass, attempting to touch the head of a child on the outside. He is apparently asking for help and companionship that he will never get.

To make things worse, he has nothing to play with except two small worn pieces of carpet. Mae Noell told me that he is 'exercised by a volunteer when one is available'.

But the exercise room was barely larger than the cage.

She also told us that they are proud of their success in breeding animals on the farm. The babies attract visitors, she said. When we asked where the babies go when they are no longer infants, we got the expected answer. Sometimes, she said, it is necessary to sell the babies to provide funds for the upkeep on the rest of the farm.

Who buys the animals? Circuses, zoos and research laboratories.

Further along the walkway is a barren cage containing a chimp named Konga. He was born in 1948 and was one of the Noells' original boxing chimps. In fact, he still has the unremovable chain around his neck to prove it.

He looks old and tired and forlorn, his face wrinkled with age, weather and stress. He begs to tourists as they pass, extending both arms outside the bars in the hope of catching a piece of monkey chow.

Next door there is Johnnie. The sign in front of his cage says he was a former 'show chimp' and that someone has taught him to speak. He very clearly enunciates the syllables 'MaMa' as he extends his hand for a peanut from a passerby.

Then he retreats again to the back of his cage where he sits and rocks endlessly, his eyes fixed perhaps on another world.

Rosie, a former 'pet', is a weary witness to what happens when an exotic animal outgrows a domestic setting. She was once someone's 'baby'. Now she is a prisoner in a dirty cell.

Chimps like Rosie often spend twenty or thirty years in solitary confinement with not one bit of mental stimulation or the touch of another being – simply because they have outgrown the age when they can be safe and cuddly human companions. Rosie's eyes tell the story of pain and loss beyond words.

As pitiful as these animals' conditions are, they are among the most presentable of the animals at the Chimp Farm. They are housed nearest the front of the facility and permitted the closest interaction with the public.

Other animals, such as Cheetah, who, the Noells claim, once performed with Johnny Weismuller of Tarzan fame, have gone insane from confinement. They are hidden from the public by opaque plastic screens. All the public can see are shadows of animals pacing endlessly back and forth.

At the back of the farm, isolated from the public by a chain-link fence, several yards of space, and high cage bars, sit several other chimps whom the Noells consider only semi-respectable.

Some are evidently insane – like Mike who was acquired from a dealer and reputedly came from an especially noxious laboratory experiment. Others are aggressive and throw faeces and food at passers-by.

Some just pace and bite themselves or rattle their cage bars. One chimp simply curls up in a foetal position and no amount of public clamour can elicit a response. She even lies there as they pelt her with food and yell insults. Her eyes are open and she is awake. But she is no longer listening.

These are just a few of the pitiful animals who are imprisoned on the Chimp Farm. Others include baboons, orang-utans, several species of monkey, gorillas, pigs and a huge bear whose cage size is in blatant defiance of both state and federal statutes.

He, and all the others, are doomed to live out the rest of their natural lives in confinement and misery.

The Chimp Farm is just one of several hundred roadside zoos across the United States.

As we were leaving the Chimp Farm, the woman we had talked to earlier in the day approached us again. I thought we were in for another lecture. Instead, she gave us a slightly embarrassed look.

'You know, you were right,' she said. 'These animals aren't happy. They look like mental patients. I won't be coming back.'

As she drove away, I thought that we were just one step closer to the day when everyone would see these places for the pitiful prisons they are.

30

They Are Us

GEZA TELEKI

In 1986 Geza Teleki was among the founders of the Committee for Conservation and Care of Chimpanzees, based in Washington, DC. He is now its chairman. Teleki has taught anthropology at George Washington University, in Washington, DC, but at present his efforts are directed toward the establishment of a new national park for chimpanzees in Sierra Leone. He is the author of The Predatory Behavior of Wild Chimpanzees *and co-editor, with Robert Harding, of* Omnivorous Primates. *In this chapter he describes how his encounters with free-living chimpanzees persuaded him to drop his original plans for a scientific career, and instead devote himself to the cause of defending the interests of chimpanzees. He describes the modern trade in chimpanzees from Africa and compares it to the slave trade. Dr Teleki reminds us that every chimpanzee is an individual with a right to freedom and self-determination.*

Two and a half decades ago, when I was a novice undergraduate at George Washington University, the study of live primates, even chimpanzees, held no interest for me. My dream then was to be a paleoanthropologist stalking the remains of early hominids somewhere in East Africa. So I, like many others of my age and era, spent my time locked within stark academic walls that narrowed my focus to primate graveyards and asylums which, back in the naive 1960s, I knew only as museums and zoos.

When I first travelled to East Africa in 1968 to study chimpanzees in the wild at Gombe National Park in Tanzania (which happened at the urging of Louis Leakey who, perhaps weary of my persistent postal pursuit of a digging career in Olduvai Gorge, unexpectedly forwarded me to Jane Goodall), my graduate studies at Pennsylvania State University nearly ended due to strong professorial objections that two years with wild chimpanzees would be a waste of time. Only the intervention of Ray Carpenter, who alone endorsed what others saw as useless diversion, kept me enrolled as a student of physical anthropology.

At Gombe, my interest in dead things waned and a stronger fascination with living beings emerged. The close associations I experienced with those chimpanzees altered my views in ways both basic and broad, leading me, albeit some years after my departure from Gombe in 1971, to shift from an aspiration to be a scientist toward a truer vocation to be a conservationist.

No single event caused that change of heart and mind, but one evening in the Gombe hills is embedded in my memory as a seed of transformation. Seeking a brief respite from months of continuous chimpanzee following, I took a rare day off to climb one of the steep ridges leading upward to the high rift escarpments that demarcated the eastern boundary of the park. As I sat alone at the crest of a grassy ridge watching a spectacular yet common sunset over the silvery waters of Lake Tanganyika in wonderful solitude and silence, I suddenly noticed two adult male chimpanzees climbing toward me on opposite slopes. They saw one another only as they topped the crest, just yards from my seat beneath a tree, whereupon both suddenly stood upright and swiftly advanced as bipeds through waist-high grass to stand close together, face to face, each extending his right hand to clasp and vigorously shake the other's while softly panting, heads bobbing. Moments later they sat down nearby and we three watched the sunset enfold the park. When dusk fell my two companions went off to build platform nests high in the trees of the valley. Nevermore, I realised as I hastened homeward to my own bed (a lower platform) at the field station before darkness fell, would I regard chimpanzees as 'mere animals'. On that singular eve, which also marked the twilight of my youth, I had seen my species inside the skin of another.

But for minor differences – a few inches in height, some extra body hair – there on a ridge in Africa had stood two colleagues performing a greeting ritual seen daily on any campus in America. Science was not a key to insight on that occasion, and anthropomorphism was for me a moot issue. The episode, and others like it later, forged a familiarity with individuals which eventually coalesced as a profound respect for a kindred species. And that, in turn, spawned a lifetime commitment to raising survival odds for all chimpanzees.

By every criterion ever devised to set humankind on the apex of species development – behavioural, social, psychological, biological – chimpanzees qualify as the closest living relatives we humans have on this planet. Not the same, inasmuch as every species by definition is distinctive, but extraordinarily similar – neighbours, so to speak, on a planetwide spectrum of diversity.

Scientific evidence confirming the close kinship can be seen in a wide range of shared traits. Example: chimpanzees exhibit many technical skills, from using stone implements to making assorted plant tools for specific purposes, which are acquired by learning and feature cultural variation. Example: chimpanzees share with us many cognitive abilities, such as long-term memory, self-recognition, sense of humour, even some elements of linguistic talent. Example: chimpanzees are so close to humans in body structure and chemistry that blood transfusions and organ transplants are feasible. Example: many emotional states are so alike for both chimpanzees and humans that each intuitively understands how to interact with the other. The list of similarities is long even if only the scientifically verifiable traits are cited; it becomes much longer if anecdotal observations are also accepted as a valid basis for comparison.

We humans commonly react with astonishment upon discovering that chimpanzees can do something we consider special to humankind. Any evidence of intelligence overlap provokes the greatest scepticism, as the uniqueness of that quality in us is our most cherished illusion. But is this defensive reaction anything more than a visible indicator of our anthropocentric view of the world? Would it not be more extraordinary if a species having broad genetic overlap with us did not perform acts or feel emotions or have thoughts akin to ours?

The prejudice we invoke to collectively degrade other species, and sometimes to selectively demean human populations, is embedded in our cultural heritage. Pejoratives such as 'vicious as a dog' and 'stupid as a baboon' reflect common themes in human societies. Intimate familiarity with individuals, human or nonhuman, is the key to shedding such prejudices as unfounded myths.

Having spent some years in the company of chimpanzees, both free and confined individuals, I find myself no longer able to cleave to the majority human view of chimpanzees as inferior beings. When I am asked today, twenty-four years after my first trip to Gombe, what drives my personal concern for chimpanzee welfare and survival, I can only reply: 'Individuals whom I know or may some day meet.' And therein lies the essence of the lesson I learned at Gombe, where each chimpanzee was special in character and also vital to community life. Knowing individuals, I cannot continue thinking in terms of abstract prejudice. My view of chimpanzees is much like my view of humans: some are scoundrels and some are saints, and most are somewhere between those extremes.

To allay critics who would say that I merely have acquired a pancentric bias, a shift of allegiance from human to chimpanzee, I should add that I view dogs in my neighbourhood much the same way because I also know them as individuals. Familiarity with the members of any group or population or species is, to me, the main antidote to indifference about the well-being and the survival of others. Had I never met a chimpanzee on equal terms, in a setting of mutual freedom, I might well have retained my ingrained anthropocentric prejudices without regret.

Much as I would prefer to believe otherwise, there is little doubt in my mind that the increasingly fragmented populations of chimpanzees that are so thinly dispersed across equatorial Africa will face extinction in the foreseeable future. The scenario being sketched by primatologists is grim at best, with land degradation and resource exploitation accelerating the demise of chimpanzee habitats and communities everywhere.

So long as humans continue to reproduce at the current rate, there can be no safe haven on the planet for other large mammals. Equatorial Africa today contains at least 300 million humans compared with less than 300 thousand chimpanzees. That thousand-fold gap widens further as humans increase at a rate of 3.1 per cent per annum and chimpanzees decrease ever more precipitously each year. If the trend continues, is it not fanciful to expect there to be room for chimpanzees anywhere in Africa by the end of the century?

In their wilderness retreats, free chimpanzees are under assault by waves of humans bearing hoes, saws and guns. Few chimpanzee communities in Africa are today safe from human encroachment and persecution. The national population estimates ring alarm bells. In the twenty-five nations encompassing the historical range of the species, four contain no chimpanzees and fifteen others retain less than 5,000 chimpanzees apiece. Survival is not assured even in the six remaining nations where populations are still relatively intact, due to recent sales of extensive timber concessions.

During the early 1980s a survey of Gabon, containing some of the best habitats, yielded an estimate of about 64,000 free chimpanzees. Biomedical scientists, whose interest in these apes has always been a consuming one, promptly cited Gabon as proof that Africa still has a 'plentiful supply' of free chimpanzees available for 'harvesting' to save human lives. But by April of 1988 the surveyors, Caroline Tutin and Michel Fernandez, stated that 'in the five years since completion of the census the situation has changed' so much that 'by 1996 the chimpanzee population of Gabon will be reduced by at least 20% as a result of habitat alteration caused by selective logging'. Other major population nucleuses in Cameroon and Zaïre are similarly threatened by rapid change.

In their prison settings, confined chimpanzees continue to suffer abusive treatment, social isolation, mental deprivation, emotional trauma and the like. I estimate that between 4,000 and 5,000 chimpanzees exist worldwide in medical institutions, zoological exhibits, roadside menageries, entertainment compounds and the homes of pet owners. Conditions of confinement may vary but it is a clear truth, in my mind, that no imprisoned chimpanzee today receives what I would regard as optimum living conditions and proper treatment.

The plight of chimpanzees in medical laboratories causes me the greatest concern. In the United States, where captive census data are most readily available, about 2,000 of 3,000 confined chimpanzees exist in biomedical facilities. It is indeed a sad statement on human values that the very institutions which proclaim a dedication to alleviating suffering and pain in humans cause so much distress to chimpanzees. And it is equally perplexing that medical scientists are the most dedicated opponents to the enacting of legislation designed to better protect free chimpanzees and improve the treatment of confined chimpanzees. I rest my case on two examples of this remarkably inconsistent position.

First, after several conservation groups petitioned the US Department of Interior's Fish and Wildlife Service to place chimpanzees on the endangered list under the Endangered Species Act, the government received 54,212 letters of support and only nine letters of dissent in 1988. Supportive letters included many from a wide range of institutions, but no biomedical facilities, while opposing letters included eight from biomedical research centres and one from a circus. Acting on behalf of the medical community, the government's National Institutes of Health mounted an intensive lobbying campaign to convince Congress that endangered status was not warranted due to presence of an 'ample supply' of chimpanzees in Africa. Because the figures collected from thirty-nine field scientists could not be easily disputed, some members of the biomedical community attacked the credibility of the field experts instead.

Second, five years after the US Senate passed amendments in 1985 requiring laboratories to provide 'a physical environment adequate to promote the psychological well-being of primates', the US Department of Agriculture proposed new regulations under the Animal Welfare Act. The delay was caused by medical opposition on an unprecedented scale. The regulations were revised many times in public reviews that produced some 12,000 letters to the US Department of Agriculture. One round of proposals published on 15 August 1990 yielded 1,372 institutional criticisms plus an uncounted number of personal objections from the medical research community. Tremendous lobbying pressure was directed at Congress to press the US Department of Agriculture into submission. Capitulation occurred by 15 February 1991. The final regulations adopted wholesale the minimum standards of maintenance and care set many years ago by the government's National Institutes of Health. For instance, the recommended cage size for permanently confining a single adult chimpanzee remained at a meagre $5 \times 5 \times 7$ feet and even infants were not guaranteed social housing.

The connecting link between freedom and confinement is the pipeline of international commerce which causes chimpanzees more suffering and trauma than the conditions they face at either end. Dealers who stop at nothing, least of all legal barriers, to pursue their greed for cash profits, and clients who shut their eyes to the most cruel and destructive methods of acquisition so as to maintain a fantasy of propriety, work together to keep this pipeline open no matter how many anti-trafficking laws are enacted.

Based on hard evidence from a wide range of sources I can confirm that at least ten chimpanzees die for every infant that survives more than a year at the final overseas destination. If a nation such as Sierra Leone, where I served for four years as a national park director and thus had access to official records, verifiably exports as many as 3,000 infants in two decades then it is an absolute fact that some 30,000 chimpanzees were exterminated from that region in the process. Were it not for the anthropocentric blinkers we put on to hide from such realities, that alone should have prompted a general indictment of the idea that chimpanzees can be legitimate objects of trade. The raw details warrant far harsher public condemnation, however, as the modern chimpanzee trade closely resembles the methods and motives of the historical slave trade.

Most prized by dealers and their clients are nursing infants who are less than two years of age and totally dependent on their mothers for survival. The slaughter begins in the wilderness as hunters with shotguns or flintlocks loaded with pebbles or metal shrapnel attack mothers and other protective group members. Many infants die when this crude ammunition scatters to hit both mothers and their clinging offspring. Pit traps, poisoned food, wire snares, nets and even dog packs are also used to kill adults defending the youngsters. More deaths occur during transport to villages. Infants are often tied hand and foot with wire, causing circulation loss and septic wounds, and are trucked to urban centres in tiny cages or tightly bound sacks, often under heavy suffocating loads to avoid detection at checkpoints. Few receive care *en route*, so starvation and dehydration are commonplace. While awaiting shipment overseas, more die of neglect in filthy holding pens and at airports where flight delays lead to exposure. Cramped in tiny crates, even carried in personal luggage, the victims often must endure days of travel through several transit points which offer ample opportunities for falsifying documentation. Some infants manage, against all odds, to survive this ordeal only to die at the final destinations from cumulative physical and psychological trauma.

If we consider that most of the 4,000–5,000 chimpanzees now held in confinement worldwide originated in West Africa, it is easy to see that 40,000 or more were exterminated there by commercial trade in about twenty years – the span of a single chimpanzee generation. When other decimating factors are added, is it surprising that the western subspecies accounts for only 10 per cent of the total number surviving in Africa?

My work on chimpanzee survival and well-being issues these past years has yielded some insights but no simple solutions to this terrible situation. I do believe, however, that every chimpanzee has rights to the freedom and the self-determination we so highly value for ourselves. And because I see chimpanzees as individuals, some of whose experiences and memories I share, I feel a moral obligation to respect the members of a kindred species.

Looking at chimpanzees from where I stand, eye to eye, not down my sharper human nose, I consider it sheer arrogance to perpetuate the anthropocentric views established by my ancestors simply because that was the collective human impulse. As Pogo once said in a memorable cartoon: 'We have met the enemy and they are us.'

VII

Epilogue

31

The Great Ape Project –
and Beyond

PAOLA CAVALIERI and PETER SINGER

Why the Project?

Aristotle refers to human slaves as 'animated property'. The phrase exactly describes the current status of nonhuman animals. Human slavery therefore presents an enlightening parallel to this situation. We shall explore this parallel in order to single out a past response to human slavery that may suggest a suitable way of responding to present-day animal slavery.

Not long ago such a parallel would have been considered outrageous. Recently, however, there has been growing recognition of the claim that a sound ethic must be free of bias or arbitrary discrimination based in favour of our own species. This recognition makes possible a more impartial appraisal of the exploitative practices that mark our civilisation.

Slavery in the ancient world has been the subject of a lively debate among historians. How did it arise? Why did it end? Was there a characteristic 'slave mode of production'? We do not need to go into all these disputes. We shall focus instead on the distinctive element of slavery: the fact that the human being becomes property in the strict sense of the term. This is sometimes referred to as 'chattel slavery' – a term that stresses the parallel between the human institution, and the ownership of animals, for the term 'cattle' is derived from 'chattel'. Slave societies are those societies of which chattel slavery is a major feature. They are relatively rare in human history. The best known examples existed in the ancient world, and in North and Central America after European colonisation.

We shall focus on slavery in classical Greece and Italy, because there is a sense in which the perspective of that period is nearer to our own. In ancient times, the idea that some human beings should be under the absolute subjection of others was so much taken for granted that it went virtually unquestioned. In this respect slavery in the classical period differs from the more recent institution of slavery in America, which was criticised from the beginning. True, Aristotle refers to some opponents of slavery, but they left so slight a trace that it is hard even to identify them; and the rare critical tones that we find over a span of many centuries, whether they come from Sophists, Stoics or Cynics, are either ambiguous or inconsequential.[1] They are confined to the abstract claim that no one was born a slave, or to the interpretation of slavery as 'a condition of the soul'. The actual practice was in fact so widely accepted and so pervasive that it has been claimed that 'There was no action or belief or institution in Graeco-Roman antiquity that was not one way or other affected by the possibility that someone involved might be a slave.'[2] There is an evident parallel here with the way in which most people still take for granted the absolute subjection of animals by human beings.

What then did it mean, for a human being, to be a piece of property in classical Athens or Rome? Obviously much depended on the particular times and circumstances in which the slave lived, but a few generalisations can be made. Slaves did not occupy any definite place in the social scale, nor in the economy. Chattel slaves could be peasants, miners, tutors, herdsmen, wet-nurses or artisans. They might live in twos and threes with a small farmer, or be part of large gangs attached to the estate of a great landowner. They might also share the urban life of an aristocratic family in whose service they were. Because of these differences, there are wide variations in opinion among historians about the extent to which the lot of the slave was a miserable one. Some have emphasised the appalling situation of the slaves working in the Greek silver mines of Laurion, from where no fewer than twenty thousand escaped during the Peloponnesian war. Equally harsh was the treatment of the chain-gangs employed to cultivate the huge farms of Sicily, where during the late second century BC rebellion was endemic. Other scholars, more prone to defend the reputation of classical civilisation, point instead to the bonds of affection that could tie the family slaves to their master, evidence of which is to be found in a number of funeral inscriptions, and in the many tales about the loyalty and affection of slaves.

Such disputes, and the wide range of possibilities that they reveal, make it difficult to grasp what the situation of the slave really was. If, however, we look behind the variety of external forms, we find one stable element: a condition of powerlessness. The treatment that the slave is accorded depends solely on the master. As a chattel, slaves have lost control over their own selves. And at the root of the master's power over the slave is the fact that the slave is not acknowledged by the community. Slave status is characterised by what the slave is not: slaves are not free, they cannot determine how to use their own labour, they cannot own property, they generally cannot testify in court, or if they can, it is usually under torture; even their family is not their own, for although family ties may be recognised in practice, slaves have no rights as parents and their children belong to their master. Bought and sold as objects, liable to corporal punishment and sexual exploitation, they stand outside the protective moral realm of the classical community.

All this confirms the parallel we suggested at the outset. As with slaves in ancient times, so with animals today, treatment varies – from the affectionate care of the 'pet-owner' to the naked exploitation of the factory farmer concerned only with maximising profits. The common thread, again, is that the animals have suffered a total loss of control over their own lives. The difference in interests and capacities of human and animal slaves does not affect their fundamental identity of status. Like chattel slaves, nonhuman animals stand outside the protective moral realm of the modern community.

Political action on behalf of animals today focuses on abolishing practices, such as the bullfight or the keeping of hens in battery cages; or on changing forms of treatment – for example, replacing hot-iron branding with less painful ways of identifying bovines, or using local anaesthetics when performing mutilating operations such as castration and tail-docking in pigs, bovines and sheep. A look at the history of slavery shows that this type of approach had only a marginal impact. Gladiatorial games were abolished, but this did not influence the general condition of slaves. Neither was this condition fundamentally changed by the undoubtedly desirable regulations that ruled out the most blatant forms of mistreatment, for example branding on the face or castration.

Slaves did, however, have one resource that animals do not have: they could rebel. This might seem to mark an important difference, but it had little effect. Although the debate on the reasons for the end of classical slavery is lively and still unresolved, no scholar regards slave rebellions as a significant factor in ending slavery.[3] The explanations offered vary, and may give prominence to economic, political, religious or sociological causes, but as far as slave rebellions are concerned, the thousands of crosses that bordered the road from Capua to Rome after the defeat of Spartacus still serve as a symbol of the normal outcome of these events. On the other hand, the few instances of briefly successful rebellions did nothing but confirm the exclusion of slaves from the community, for they left behind bands of outlaws who survived at the boundaries of inhabited areas, more or less as feral animals do in the modern world.

In antiquity there was one way to leave the no man's land of the slave: manumission. The term, which means literally to emit or release from one's hand, vividly conveys the sense of giving up control over something. It was, once again, the unilateral act of the master. By giving up his dominion through an act that could be religious or civil, formal or informal, the master sanctioned for the slave the end of her or his condition as property and the beginning of some form of recognition within the moral community.[4] In republican Rome the change was even more radical: together with freedom, the slave received citizenship, thus passing in one step from the status of absolute outsider to that of full member of the dominant group.

One objection that has frequently been raised against the idea of animal liberation is that animals cannot fight for themselves. They can only be liberated by others. Yet in this respect too the parallel with ancient slavery holds in practice, because slave rebellions had so little effect on the institution as a whole. Moreover, just as manumission was the only way out for slaves, so it appears to be the kind of response needed for animals. Not only does it address directly the question of status; it is also a long-standing and well-known instrument for the bestowal of freedom on those who cannot win it for themselves. When applied to animals, moreover, manumission has a further advantage. Though it cannot be used for all animals at once (for practical reasons, and also because it is of its nature a reformist measure) it can still create a precedent. As a tool for systematic intervention, each use invites us to consider the possibility of applying the tool in another situation.

Some problems remain. Even when manumission was applied to large groups of slaves, the freeing of each slave was, legally speaking, an individual event. What is needed for animals seems instead to be a distinct, more symbolic grant of freedom and moral status. Did this ever occur in Greco-Roman antiquity? There is at least one instance of this, although it concerns not chattel slavery, but a different institution, sometimes referred to as 'collective bondage'. Sparta's helots are the best-known example. Both the origins and the details of the status of the helots are somewhat obscure and still under discussion, but helots appear to have been the class of producers on the exploitation of whom the highly hierarchical Spartan society was based. The helots were employed mainly in agriculture, and so normally lived apart from the ruling group and were allowed to have a family and some sort of community life. Their situation, however, wasn't much better than that of chattel slaves. (It is enough to note that their subjection was annually reaffirmed through ritual activities that differed from modern sport hunting mainly in the fact that the prey was human.[5])

Because the helots lived together, they were more dangerous to their masters, and many helot rebellions shook Spartan society. In 369 BC, however, during a war between the Spartans and the Thebans, something exceptional happened: after defeating the Spartans at Leuctra, Epaminondas, leader of the Theban forces, manumitted *en bloc* the helots of Messenia. This act allowed the rebirth of a whole people, with the ready return of its diaspora from all over the Greek Mediterranean; it also had the effect of transforming the ancient social structure on which Sparta had been based. In the hands of the victorious Thebans, a *collective* manumission had become a political instrument with momentous consequences.

The impact that the first nonhuman manumission could have would be much greater. The same holds, however, for the difficulties that beset it. To appreciate this, it is enough to recall that during the 2,300 years that have elapsed since Epaminondas' gesture, forms of human bondage have persisted around the world, reaching their peak in the slave societies of the New World, and coming to their official end as recently as 1962, in the Arabian Peninsula.[6] If it has taken so long to make a reality of the idea that no member of our species can be an item of property, to bestow freedom and equality on members of a species other than our own will seem an arduous and improbable enterprise.

Granted, this enterprise of expanding the moral community beyond the species boundary has on its side the power of the rational challenge that, starting with the Enlightenment, if not before, has provided the theoretical basis for so many struggles for justice and has undermined all the attempted justifications put up in defence of the exclusion of some human beings from the moral community. Indeed, the need to push the egalitarian stance beyond the boundaries of the human species appears to be built into the Enlightenment dream of a universal rationality. But this alone is not enough to ensure the success of the enterprise. The history of other social movements shows that we require a conscious strategy to achieve what Darwin refers to as 'the gradual illumination of men's minds'.[7] For those who aim at change, it is vital to understand the framework in which one has to act, and to take advantage of the contradictions in the positions of one's opponents.

Why the Great Apes?

A solid barrier serves to keep nonhumans outside the protective moral realm of our community. By virtue of this barrier, in the influential words of Thomas Aquinas, 'it is not wrong for man to use them, either by killing or in any other way whatever'.[8] Does this barrier have a weak link on which we can concentrate our efforts? Is there any grey area where the certainties of human chauvinism begin to fade and an uneasy ambivalence makes recourse to a collective animal manumission politically feasible? As the philosophers, zoologists, ethologists, anthropologists, lawyers, psychologists, educationalists and other scholars who have chosen to support this project show, this grey area does exist. It is the sphere that includes the branches closest to us in the evolutionary tree. In the case of the other great apes, the chimpanzee, the gorilla and the orang-utan, some of the notions used to restrict equality and other moral privileges to human beings instead of extending them to all sentient creatures can cut the other way. When radical enfranchisement is being demanded for our fellow apes, the very arguments usually offered to defend the special moral status of human beings *vis-à-vis* nonhuman animals – arguments based on biological bondedness or, more significant still, on the possession of some specific characteristics or abilities – can be turned against the status quo.

Chimpanzees, gorillas and orang-utans occupy a particular position from another perspective, too. The appearance of apes who can communicate in a human language marks a turning-point in human/animal relationships. Granted, Washoe, Loulis, Koko, Michael, Chantek and all their fellow great apes cannot directly demand their general enfranchisement – although they can demand to be let out of their cages, as Washoe once did[9] – but they can convey to us, in more detail than any nonhuman animals have ever done before, a nonhuman viewpoint on the world. This viewpoint can no longer be dismissed. Its bearers have unwittingly become a vanguard, not only for their own kin, but for all nonhuman animals.

Nevertheless, it might be said that in focusing on beings as richly endowed as the great apes we are setting too high a standard for admission to the community of equals, and in so doing we could preclude, or make more difficult, any further progress for animals whose endowments are less like our own. No standard, however, can be fixed forever. 'The notion of equality is a tool for rectifying injustices . . . As is often necessary for reform, it works on a limited scale.'[10] Reformers can only start from a given situation, and work from there; once they have made some gains, their next starting-point will be a little further advanced, and when they are strong enough they can bring pressure to bear from that point.

How can we advocate inclusion or exclusion for whole species, when the whole approach of animal liberationist ethics has been to deny the validity of species boundaries, and to emphasise the overlap in characteristics between members of our own species and members of other species? Have we not always said that the boundary of species is a morally irrelevant distinction, based on mere biological data? Are we not in danger of reverting to a new form of speciesism?

This is a problem that has to do with boundaries, and boundaries are here tied up with the *collective* feature of the proposed manumission. What, then, can be said in favour of such a collective manumission, apart from recognition of its obvious symbolic value? We think that a direction can be found, once more, in history. It is already clear that in classical antiquity, while the collective emancipation of Messenian helots led to some dramatic social changes, the random manumission of individual human slaves never led to any noteworthy social progress. Even in more recent times, when a conscious political design was not only feasible, but was actually pursued – namely, during the first stages of the anti-slavery struggle in nineteenth-century United States – the freeing of individual slaves, or even the setting free by an enlightened plantation owner of all the slaves on his plantation, did little good for the anti-slavery side as a whole. Given that the global admission of nonhuman animals to the community of equals seems out of the question for the moment, one way to avoid a parallel failure is to focus on the species as a collectivity, and to opt for (otherwise questionable) rigid boundaries.

Finally, we cannot ignore doubts about the practical feasibility of the project, and the concrete implications of admitting chimpanzees, gorillas and orang-utans into the community of equals. Quite novel problems are likely to arise, but they will not be insuperable, and as we overcome each one, we will reveal the spurious nature of the alleged obstacles to overcoming the boundaries between species. In fact, difficulties also occurred in similar situations involving humans, but this was no reason to abandon the overall plan of emancipation. Readers will not need to be reminded that the liberation of the American slaves after the Civil War was not sufficient to achieve equal civil rights for them. Instead, a new set of obstacles to equality arose, some of which were overcome only by the civil rights movement of the 1960s, while others remain a problem today.

For the idea of providing a restitution of orang-utans, gorillas and chimpanzees to their lands of origin, in particular, we can even identify a precise historical antecedent: the creation in Africa of the state of Liberia, which the American colonisation movement dreamt would be a new homeland for those humans who had been enslaved and transported across the ocean by other members of their species. The fact that an independent nation called Liberia still exists shows that the project was feasible, and if there were things that went wrong, they were, significantly enough, related to typically human questions, for example, the discrimination against the native inhabitants of the area that the immigrants soon practised.[11]

Certainly, with regard to nonhumans, one can point to additional problems. They will not be able to participate in the political structure of the community. Unlike the descendants of the American slaves, they will be unable to stand up in defence of their own rights, or of any territory that they are granted, whether in Africa or in other countries as well. How then is it going to be possible to ensure for them the same protection afforded to full members of our community? Here the analogy with the emancipation of slaves breaks down; but to some extent we can draw on two further models, depending on the kind of situation in which the liberated apes will be. If they are living in natural conditions in their own territories, whether in their homelands or in the countries to which they have been coercively transported, they will have no need of our assistance; they will need only to be left alone. We do have world institutions – though imperfect ones – that exist in order to protect weaker countries from stronger ones. We also have considerable historical experience with the United Nations acting as a protector of non-autonomous human regions, known as United Nations Trust Territories. It is to an international body of this kind that the defence of the first nonhuman independent territories and a role in the regulation of mixed human and nonhuman territories could be entrusted.

Where, on the other hand, individuals have become so habituated to life within our societies that it would not be in their own interests to be returned to wild habitats, their status, and the protection to be afforded them, could be just the same as that which we grant to non-autonomous beings of our own species. As in the case of children and the intellectually disabled, the basic protection ensured by national laws could be supplemented by specially appointed guardians. In fact, it is not only animals and non-autonomous humans who are unable to stand up in defence of their own rights; often normal adult humans have been in need of protection. This is the *raison d'être* of many international organisations, such as the venerable Anti-Slavery Society for the Protection of Human Rights or the Fédération International des Droits de L'Homme, created after the Dreyfus Affair, or the more recently founded Amnesty International, an organisation that makes the 'guardian' nature of its work clear by speaking of the 'adoption' of a political prisoner, when a local group takes up that prisoner's cause. These nongovernmental organisations oversee, within the constraints of their moral influence and any political power or sanctions that they are able to wield, the realisation of the various international declarations of human rights in signatory nations. Their work is further evidence for the necessity of creating the world institution that we have envisaged. By combining some of the functions of existing models, such a body could be in a position to carry on the complex task of monitoring the implementation of a declaration of rights for the great apes – wherever they may be.

The creation of such an international body for the extension of the moral community to all great apes will not be an easy task. If it can be accomplished it will have an immediate practical value for chimpanzees, gorillas and orang-utans all over the world. Perhaps even more significant, however, will be its symbolic value as a concrete representation of the first breach in the species barrier.

Notes

1. See Giuseppe Cambiano, 'Aristotle and the anonymous opponents of slavery', in Moses I. Finley (ed.), *Classical Slavery* (Frank Cass, London, 1987); see also Robert Schlaifer, 'Greek theories of slavery from Homer to Aristotle', in Moses I. Finley (ed.), *Slavery in Classical Antiquity* (Heffer,

Cambridge, 1960), in particular pp. 199–201; and, for a general survey of the voices of protest, Zvi Yavetz, *Slaves and Slavery in Ancient Rome* (Transaction Books, New Brunswick, NJ, 1988), pp. 115–18.

2. Moses I. Finley, *Ancient Slavery and Modern Ideology* (Penguin, Harmondsworth, 1980), p. 65.

3. For the debate about the end of classical slavery, see, for example, Finley, *Ancient Slavery*, pp. 11–66; Yvon Garlan, *Les esclaves en Grèce ancienne* (Editions La Découverte, Paris, 1984), pp. 13–26: Jean Christian Dumont, *Servus: Rome et l'esclavage sous la République* (Ecole Française de Rome, Rome, 1987), pp. 1–20; Zvi Yavetz, *Slaves and Slavery*, pp. 115–53.

4. On manumission in Greece, see Aristide Calderini, *La manomissione e la condizione dei liberti in Grecia* (Hoepli, Milan, 1908); and for Rome, W.W. Buckland, *The Roman Law of Slavery* (Cambridge University Press, Cambridge, 1908), Part II. For more recent surveys of manumission in Greece and Rome, see William L. Westermann, *The Slave Systems of Greek and Roman Antiquity* (American Philosphical Society, Philadelphia, 1955), *passim* and, in particular, pp. 18ff., 89–90; and Thomas Wiedemann, *Greek and Roman Slavery* (Croom Helm, London, 1981), Ch. 3.

5. On this institution, called 'krypteia', see the essay by Pierre Vidal-Naquet, 'Le chasseur noir et l'origine de l'éphébie athénienne', in P. Vidal-Naquet, *Le chasseur noir: Formes de pensée et formes de société dans le mond grec* (Librairie François Maspéro, Paris, 1981).

6. 'Slavery in Saudi Arabia ended by Faisal edict', *New York Times*, 7 November 1962. See also the answer of Saudi Arabia to the first question of the UN Questionnaire on Slavery, in Mohamed Awad, *Rapport sur l'esclavage* (United Nations, New York, 1967), pp. 126–9.

7. From a letter by Darwin, supposedly written to Karl Marx, but probably to Marx's son-in-law, Edward Aveling, published in Erhard Lucas, 'Marx und Engels: Auseinandersetzung mit Darwin zur Differenz zwischen Marx und Engels', *International Review of Social History*, vol. 9 (1964) pp. 433–69, quoted in James Rachels, *Created from Animals* (Oxford University Press, Oxford, 1990), p. 101.

8. Thomas Aquinas, *Summa Contra Gentiles* (Benziger Brothers, Chicago, 1928), Book 3, Part II, Ch. CXII, reprinted in T. Regan and P. Singer (eds), *Animal Rights and Human Obligations*, 2nd edn (Prentice-Hall, Englewood Cliffs, NJ, 1989), p. 8.

9. Eugene Linden, *Silent Partners* (Times Books, New York, 1986), p. 30.

10. Mary Midgley, *Animals and Why They Matter* (University of Georgia Press, Athens, GA, 1983), p. 67.

11. On the differences in the way in which the 'Americo-Liberians' and the indigenous inhabitants of Liberia were treated, see F.J. Pedler, *West Africa* (Methuen, Birkenhead, 1959), pp. 134–6; and C.S. Clapham, *Liberia and Sierra Leone* (Cambridge University Press, Cambridge, 1976), pp. 8–9, 17ff.